Metabolomics - Methodology and Applications in Medical Sciences and Life Sciences

Edited by Xianquan Zhan

Published in London, United Kingdom

IntechOpen

Supporting open minds since 2005

Metabolomics - Methodology and Applications in Medical Sciences and Life Sciences
http://dx.doi.org/10.5772/intechopen.90987
Edited by Xianquan Zhan

Contributors
Anna Maria Timperio, Veronica Lelli, Antonio Belardo, Chiara Carazzone, Julie P.G. Rodriguez, Mabel Gonzalez, Gerson-Dirceu López, Dr. Jeyapragash Danaraj, Saravanakumar Ayyappan, Yosuva Mariasingarayan, Maria F. Shishova, Vladislav V. Yemelyanov, Roman K. Puzanskiy, Mikhail S. Burlakovskiy, Ludmila A. Lutova, Abdul-Hamid Emwas, Kacper Szczepski, Ryan McKay, Hiba Asfour, Chung-Ke Chang, Joanna Lachowicz, Mariusz Jaremko, Katerina Sazanova, Nadezhda Psurtseva, Alexey Shavarda, Xianquan Zhan, Jingru Yang, Shu Zheng, Nannan Li, Na Li

First published in London, United Kingdom, 2021 by IntechOpen
IntechOpen is the global imprint of INTECHOPEN LIMITED, registered in England and Wales, registration number: 11086078, 5 Princes Gate Court, London, SW7 2QJ, United Kingdom
Printed in Croatia

British Library Cataloguing-in-Publication Data
A catalogue record for this book is available from the British Library

Additional hard and PDF copies can be obtained from orders@intechopen.com

Metabolomics - Methodology and Applications in Medical Sciences and Life Sciences
Edited by Xianquan Zhan
p. cm.
Print ISBN 978-1-83969-083-9
Online ISBN 978-1-83969-084-6
eBook (PDF) ISBN 978-1-83969-085-3

For EU product safety concerns:
IN TECH d.o.o., Prolaz Marije Krucifikse Kozulić 3, 51000 Rijeka, Croatia, info@intechopen.com or visit our website at intechopen.com.

Meet the editor

Xianquan Zhan received his MD and Ph.D. in Preventive Medicine from the West China University of Medical Sciences. He received his post-doctoral training in oncology and cancer proteomics at Central South University, China, and the University of Tennessee Health Science Center (UTHSC), USA. He worked at UTHSC and the Cleveland Clinic during 2001-2012, and achieved the rank of associate professor at UTHSC. Currently, he is a full professor at Central South University and Shandong First Medical University, as well as an advisor to MS/Ph.D. graduate students and postdoctoral fellows. Dr. Zhan is a fellow of the Royal Society of Medicine and the European Association for Predictive, Preventive & Personalized Medicine (EPMA), a national representative for EPMA, and a member of the American Society of Clinical Oncology (ASCO) and the American Association for the Advancement of Sciences (AAAS). He is editor in chief of the *International Journal of Chronic Diseases & Therapy*, associate editor of *EPMA Journal* and *BMC Medical Genomics*, and guest editor of *Frontiers in Endocrinology*, *Mass Spectrometry Reviews*, *EPMA Journal*, and *Oxidative Medicine and Cellular Longevity*. He has published more than 140 articles, twenty-five book chapters, five books, and two US patents in the field of clinical proteomics and biomarkers.

Contents

Preface

The center of multiomics is moving from genome-centered studies to phenome-centered studies in life sciences and medical sciences. Metabolomics, as the important aspect of phenomics, is the methodology and theory for studying the metabolome, including identifying its biochemical and molecular characteristics, characterizing interactions among different metabolites or between metabolites and genetic/environmental factors, and evaluating biochemical mechanisms related to given conditions such as different pathophysiological processes. The metabolome contains all metabolites derived from sugars, lipids, proteins, and nucleic acids in a given biological system, tissue, cell, or body fluid. The metabolites in a metabolome interact mutually in enzymatic reaction systems to form metabolic network systems. Alteration in the metabolome is associated with multiple factors, including genetic, environmental, internal, external, drug, or dietary factors. Currently, the studies on variations in the metabolome are insufficient. Metabolomics includes a targeted approach based on selected/multiple reaction monitoring (SRM/MRM) and an untargeted approach based on nuclear magnetic resonance (NMR) or mass spectrometry (MS). The development of high-throughput, high-sensitivity, and especially high-reproducibility approaches is necessary to maximize the coverage of variations in the metabolome. Studies on metabolomic variations result in the discovery of effective biomarkers to clarify molecular mechanisms of a disease and determine reliable therapeutic targets. These studies also benefit precise prediction, diagnosis, and prognostic assessment in the context of predictive, preventive, and personalized (3P) medicine, as well as contribute to life sciences.

This book focuses on the advancements in the concept and methodology of metabolomics, including sample preparation, targeted metabolomics based on SRM/MRM, untargeted metabolomics based on NMR or MS, and applications of metabolomics in the research and practice of medical sciences and life sciences. Chapter 1 addresses the concept and importance of metabolomics and its methodology, including targeted and untargeted approaches. Chapter 2 addresses the versatility of metabolomics analysis due to the enormous variety of samples and the absence of strict barriers between quantitative and qualitative analyses. The chapter also highlights untargeted metabolomics using the metabolomics of opposite Antarctic cryptoendolytic communities as an example, and targeted metabolomics using urine during childbirth as an example. Chapter 3 suggests that pharmacometabolomics is a new horizon in personalized medicine, highlighting potentials and advantages in designing innovative and personalized drug treatment. Chapter 4 examines the theory and methodology of volatiomics of natural products from plants, flowers, fruits, microorganisms, and animals. Chapter 5 presents new insights into seagrass metabolomics in marine-based drug discovery through the study of its bioactive products. Chapter 6 discusses the metabolic profiling of transgenic tobacco plants that synthesize bovine interferon-gamma for optimizing the quality and quantity of plant-synthesized interferon-gamma. Chapter 7 discusses the use of gas chromatography (GC)-MS-based metabolomic changes in wood-inhabiting filamentous fungi during ontogenesis.

This book contains only a fraction of the very important metabolomic studies in medical sciences and life sciences. However, the studies included serve to stimulate and encourage researchers to study and practice metabolomics, especially to discover effective biomarkers, novel therapeutic targets/drugs, and patient stratification for diseases. Moreover, the extensive studies on metabolomics in life sciences, in turn, promote and serve research and clinical practice in medicine.

Xianquan Zhan, MD, PhD, FRSM
Professor of cancer proteomics and PPPM,
Medical Science and Technology Innovation Center,
Shandong First Medical University,
China

Chapter 1

Introductory Chapter: Metabolomics

Xianquan Zhan, Jingru Yang, Shu Zheng, Nannan Li and Na Li

1. Introduction

The center of multiomics is being moved from genomics to phenomics (**Figure 1**) [1]. Proteome and metabolome are two main components of phenome, and are equally important. The concept and development of proteoforms significantly enrich the content of a proteome. A book entitled "Proteoforms: Concept and Applications in Medical Sciences" has been published focusing on proteomics at the proteoform level [3]. It is driving the editor to edit another book focusing on metabolomics to discuss (i) the methodology of metabolomics, including sample preparation, targeted metabolomics, and untargeted metabolomics based on nuclear magnetic resonance (NMR) or mass spectrometry (MS), and (ii) applications of metabolomics in the research and practice of life science and medical science.

Metabolomics is an important aspect of phenomics, which is the theory and methodology to study metabolome, including identification of biochemical and molecular characteristics of metabolome, characterization of interactions among

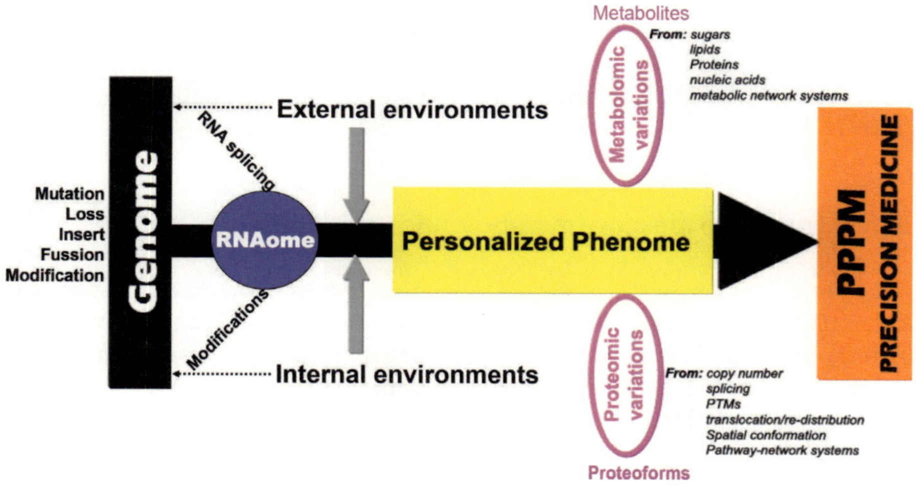

Figure 1.
The imbalance contribution of multiomics to clinical practice. RNAome includes messenger RNAs (mRNAs) and non-coding RNAs (ncRNA). Multiple modifications extensively occur at three different levels of DNAs, RNAs, and proteins to systematically regulate physiological and pathological processes. The center of multiomics is being moved from genomics to phenomics, especially proteomics and metabolomics. PTMs = post-translational modifications. PPPM = predictive, preventive and personalized medicine (3P medicine). Modified and upgraded from Zhan et al. [1], with permission from Elsevier publisher, copyright 2018; and reproduced from Li et al. [2], with permission from Wiley publisher, copyright 2021.

different metabolites or between metabolites and genetic/environmental factors, and evaluation of biochemical mechanisms related to a given condition such as different pathophysiological processes [1]. Metabolome contains all metabolites derived from nucleic acids, proteins, lipids, and sugars in a given cell, tissue, biological system, or body-fluid. The metabolites in a metabolome interact mutually in enzymatic reaction systems to form metabolic network systems. The metabolomic variation is associated with multiple factors, including genetic, environmental, internal, external, drug, or dietary factors [1]. Currently the studies on metabolomic variations are much insufficient in the width and depth of metabolomics. It is necessary to develop high-sensitivity, high-throughput, and high-reproducibility methodology for maximizing the coverage of metabolomic variations. The studies on metabolomic variations directly result in the discovery of effective biomarkers to clarify molecular mechanisms of a disease, determine reliable therapeutic targets, and discover reliable biomarkers for precise prediction, diagnosis, and prognostic assessment in the context of predictive, preventive and personalized medicine (3P medicine, PPPM).

2. Importance of metabolomic variations in medical science

Metabolome contains all metabolites derived from nucleic acids, proteins, lipids, and sugars in a given cell, tissue, biological system, or body-fluid [4–6]. The metabolites in a metabolome interact mutually in enzymatic reaction systems to form metabolic network systems [5]. The change of metabolites is associated with multiple factors, including internal, external, genetic, environmental, drug, or dietary factors. Metabolomics is the theory and methodology to study metabolome, including identifying biochemical and molecular characteristics of metabolome, characterizing interactions among different metabolites or between metabolites and genetic/environmental factors, and evaluating biochemical mechanisms related to a given condition such as different pathophysiological processes [7]. Metabolomic variations can reflect the status of physiological and pathological processes, monitor the progression of a disease, and predict and assess the drug effects compared to the baseline of metabolic profiles, which benefits for disease stratification, and personalized/precise medicine in the context of PPPM [8].

3. Samples used to measure metabolomic variations

The biological samples are very intricate that are used to measure metabolomic variations, including extracts from different cells, tissues, and body-fluids (**Table 1**). Urine and serum/plasma [6, 17, 18] are the most commonly used body-fluids to analyze metabolome for different diseases because these samples are very easily available and are easy to be prepared, without any injury. In addition, tears [19] are the good samples for analyzing metabolome of an eye disease, exhaled air [20, 21] for pulmonary and airway diseases or other diseases, saliva [22] for oral diseases, synovial fluid [23] for arthritis, and cerebrospinal fluid (CSF) [24] for neurological systems disease. Generally speaking, there are many biological samples that are suitable for metabolomics analysis of a disease. The metabolomics studies based on these different samples can directly or indirectly reflect the status of a disease, which may use to understand the molecular mechanism of a disease, and discover therapeutic targets and reliable biomarkers to predict, diagnose, and prognostically evaluate a disease.

Biological sample	Methods	Main results	References
HeLa cells	Gas cluster ion beam-secondary ion MS (GCIB-SIMS)	Purinosomes comprise nine enzymes that act synergistically, channeling the pathway intermediates to synthesize purine nucleotides, increasing the pathway flux, and influencing the adenosine monophosphate/guanosine monophosphate ratio.	[9]
Carcinoma and adjacent normal tissues	UHPLC-Orbitrap MS	This method enables targeted profiling of over 400 biologically important metabolites covering 92 metabolic pathways	[10]
Sweat	GC–MS and LC–MS/MS	As most of the identified metabolites are involved in key biochemical pathways, this study opens interesting possibilities to the use of dry sweat as a source of metabolite markers for specific disorders.	[11]
Urine and plasma	HPLC-ESI-qTOF-MS	A total of 31 and 38 metabolites in plasma and urine, respectively, showed significant differences between healthy volunteers and Sjögren's Syndrome patients and were proposed for their identification.	[12]
Cerebrospinal fluid (CSF)	GC–MS and LC–MS/MS	A total of 274 CSF-derived metabolites were common to the discovery and replication cohorts in cancer-related fatigue.	[13]
Saliva	UHPLC-qTOF-MS	The study identified and classified a total of 211 endogenous and exogenous salivary metabolites. The results reveal a distinct metabolite profile of dog and human saliva as 25 lipid compounds were identified only in canine saliva and eight dipeptides only in human saliva.	[14]
Sputum	LC–MS/MS	The KEGG analysis revealed that the glycerophospholipid metabolism pathway was downregulated in severe COPD. Due to the critical role of glycerophospholipid metabolism in oxidative stress, significant negative correlations were discovered between glycerophospholipid metabolites and three oxidative stress products (SOD, MPO, and 8-iso-PGF2α). The diagnostic values of SOD, MPO, and 8-iso-PGF2α in induced sputum were found to exhibit high sensitivities and specificities in the prediction of COPD severity.	[15]
Blood	1D 1H NMR spectroscopy methods	This has led to the absolute quantitation of nearly 70 metabolites in serum and plasma and nearly 80 in whole blood.	[16]

Table 1.
Examples of different types of biological samples used for metabolomics analysis.

4. Methods used to measure metabolomic variations

The appropriate analytical methods for metabolomics are important to detect, identify, and quantify metabolomic variations in a given condition; for example, a disease status versus control, which are mainly classified into targeted metabolomics [25] and untargeted metabolomics [26]. (i) The targeted metabolomics [25] is to mainly quantify hypothesis-driven known metabolite variations in a metabolome (such as metabolites derived from one or more unknown metabolism pathways) between or among research groups, and then use multivariate statistical analysis to establish mathematical models [27]. This mathematical model then is used to discriminate Diseases from healthy controls, treatment from untreatment, or different stages of diseases. The often used methods for targeted metabolomics are the selected/multiple reaction monitoring (SRM/MRM) analysis with an optimized sample extraction and liquid chromatography-mass spectrometry (LC–MS) conditions using the triple quadrupole mass spectrometry (QqQ-MS) [28]. (ii) The untargeted metabolomics [26] is an none hypothesis-driven approach to globally detect, identify, and quantify metabolite variations in a metabolome in a biological system without any bias, which will benefit the understanding molecular mechanism of a disease, discover new therapeutic targets/drugs and metabolite biomarkers for effective prediction, diagnosis, and prognosis. The often used methods for untargeted metabolomics are the mass spectrometry (MS)-based methods [6, 29], and nuclear magnetic resonance (NMR)-based methods [30, 31] (**Figure 2**). (a) MS-based methods have ion mobility coupled with MS (IM-MS) that can measure time, mass-to-charge (m/z) and intensity variables [1], capillary electrophoresis coupled with MS (CE-MS) that can measure time, m/z and intensity variables [29, 32, 33], gas chromatography coupled with MS (GC–MS) that can measure time, m/z and intensity variables [29, 34], liquid chromatography coupled with MS (LC–MS) that can measure retention time (RT), m/z and intensity variables [26, 29, 35], and direct injection coupled with MS (DI-MS) that can measure m/z and intensity variables [1]. IM-MS is to use a buffer gas and a uniform or periodic electric field for separation of ions based on size and shape of the ions, followed by MS analysis. This is a very high throughput and high selectivity method, which can easily separate isomeric and isobaric compounds. CE-MS is to use electro kinetics for

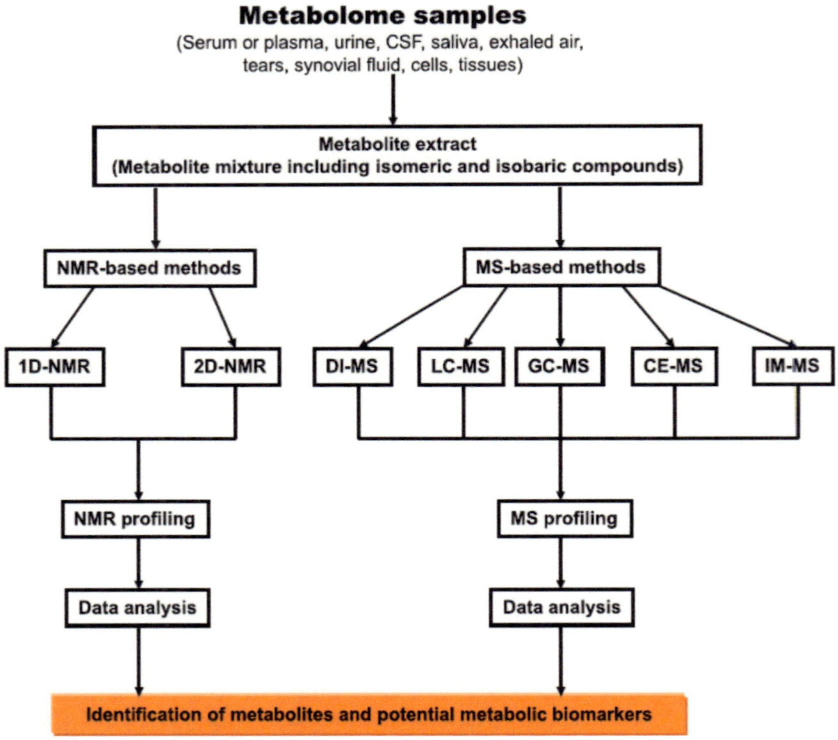

Figure 2.
The main metabolomic strategies for identification of metabolite profiling and discovery of biomarkers. Reproduced from Zhan et al. [1], with permission from Elsevier Publisher, copyright 2018.

separation of polar molecules, followed by MS analysis. This is a very good method to analyze polar molecules in aqueous samples for measurement of inorganic and organic anions, with low running costs and relatively low throughput. GC–MS is to use gas chromatography for separation of molecules, followed by MS analysis. This method is suitable for a polar and volatiles compounds, whose advantages are availability of universal database for identification, high sensitivity, and high reproducibility; and whose disadvantages are only detection of a polar and volatile compounds, requirement of derivatization of polar compounds, low ionization discrimination, and requirement of higher amount of samples. LC–MS is to use liquid chromatography for separation of molecules, followed by MS analysis. This method is suitable for polar to hydrophobic compounds, whose advantages are requirement of minimal amount of samples, high sensitivity, high throughput, and flexibility in column chemistry widening the range of detectable compounds; and whose disadvantages are requirement of high ionization discrimination, lack of large metabolite databases, and requirement of specific chromatographic conditions for very polar molecules. DI-MS is to use the nanospray source directly coupled with MS, which does not require chromatography separation, whose advantages are low sample volume requirement, high sensitivity, high-throughput, and low cost; and whose disadvantages are requirement of high ionization discrimination, significant ion suppression phenomenon, and inability to separate isomers and isbaric species. (b) NMR-based methods have one-dimensional, two-dimensional, and three-dimensional NMR methods (1D-NMR, 2D-NMR, and 3D-NMR) [31], which is to use the interaction of spin active nuclei (13C, 1H, 31P, 19F) in the electromagnetic fields for obtaining structural, chemical, and molecular environment information [30, 31], whose advantages are non-destruction of sample, minimal sample preparation, high reproducibility, relative high throughput, availability of molecular dynamic and compartmental information with diffusional methods, and availability of databases; and whose disadvantages are low sensitivity, overlapping of metabolites, and high instrumentation cost [36]. MS-based methods and NMR-based methods are complimentary for metabolomics analysis, and both will produce very complex data. The processing, analysis, and annotation of data are very important and crucial steps to discover the potential and important metabolic biomarkers [37, 38]. However, compared to the NMR-based metabolomics, MS-based metabolomics has a relatively low cost, high sensitivity and resolution, and very good analytical performance to measure the metabolomic variations for PPPM or PM practice [39].

5. Applications of metabolomics in life science and medical science

Metabolome is the important content of phenome. Metabolomics conducts qualitative and quantitative analysis of all small molecule metabolites in organisms, and searches for the relative relationship between metabolites and physiological and pathological changes. The subjects are mostly small molecules with molecular weights of less than 1,000. With the development of high throughput technology, the study of living organisms has developed from single small molecule to multi-omics; such as genomics, transcriptomics, proteomics, metabolomics. Multiomics reflects molecular changes in a disease or biological process, and molecules that can be identified can be used as valuable biomarkers. Metabolites are substances produced or consumed through the metabolic process. Metabolites are the final expression products subject to genetic control and environmental influence. Imprints with genomic, transcriptomic, epigenetic and environmental effects are called "associations between genotypes and phenotypes" [40]. Metabolomics has been extensively applied in fields of medical science and life science (**Table 2**). It has important applications in medicine and life sciences, agriculture, food safety and so on. Metabolites,

Metabolomics methods	Biological samples	Main discoveries	References
1D-NMR	Plasma samples from SARS-CoV-2 rRT-PCR-positive patients (n = 15, with multiple sampling timepoints) and age-matched healthy controls (n = 34, confirmed rRT-PCR negative), together with patients with COVID-19/ influenza-like clinical symptoms who tested SARS-CoV-2 negative (n = 35).	The study observed four plasma cytokine clusters that expressed complex differential statistical relationships with multiple lipoproteins and metabolites. These included the following: cluster 1, comprising MIP-1β, SDF-1α, IL-22, and IL-1α, which correlated with multiple increased LDL and VLDL subfractions; cluster 2, including IL-10 and IL-17A, which was only weakly linked to the lipoprotein profile; cluster 3, which included IL-8 and MCP-1 and were inversely correlated with multiple lipoproteins. IL-18, IL-6, and IFN-γ together with IP-10 and RANTES exhibited strong positive correlations with LDL1–4 subfractions and negative correlations with multiple HDL subfractions.	[41]
2D-NMR	Different aging regimes (crust from dry-aged beef, inner edible flesh of dry-aged beef, and wet-aged beef striploin)	NMR-based multivariable analyses could be used to distinguish the method, degree, and doneness of beef aging.	[42]
3D-NMR	*E. coli* cell lysate	For 19 of the 25 model metabolites, "Structure of unknown metabolomic mixture components by MS/NMR" yielded complete structures that matched those in the mixture independent of database information.	[43]
DI-MS	A parasite–host cell system	The study applied a metabolic fingerprinting approach to evaluate metabolic changes induced by six different (candidate) drugs in a parasite–host cell system.	[44]
LC–MS	*P. aeruginosa* (35 clinical strains)	Those clinical strains that differed in their virulence and biofilm phenotype also had pronounced divergence in their metabolomes, as underlined by 332 features that were significantly differentially abundant with fold changes greater than 1.5 in both directions.	[45]
GC–MS	Embryonic zebrafish	A total of 87 important endogenous metabolites such as citric acid and hypoxanthine were identified by universal databases or standards among 270 extracted metabolites, which consisted of sugars, amines, amino acids, nucleotides, fatty acids, and sterols.	[46]
CE-MS	Plasma samples of acute corneal seizure mouse model.	Both electrically induced seizures showed decreased values of methionine, lysine, glycine, phenylalanine, citrulline, 3-methyladenine and histidine in mice plasma. However, a second provoked seizure, 13 days later, showed a less pronounced decrease of the mean concentrations of these plasma metabolites, demonstrated by higher fold change ratios.	[47]

Metabolomics methods	Biological samples	Main discoveries	References
IM-MS	Breast cancer plasma samples	Analysis of the resulting data showed that phosphatidylcholines, triglycerides and diglycerides exhibited lower expression and phosphatidylserine showed increased expression in the breast cancer samples compared to those of healthy subjects. The coefficients of variation, determined by reference to the QC data, for all of the features identified as potential markers of disease, were 6% or less.	[48]

Table 2.
Examples of different metabolomics applied in life science and medical science.

as the end products of gene expression, have been implicated in many diseases. For example, metabolomics has great potential for diabetes research, metabolic markers hold the potential to detect diabetes-related complications already under subclinical conditions in the general population [49]. Metabolomics is used to identify key disease-related metabolic changes and disease-progression-related changes, and defining metabolic changes during AD disease trajectory and its relationship to clinical phenotypes provided a powerful roadmap for drug and biomarker discovery [50]. Carmen Peña-Bautista's work shows that the untargeted analysis carried out in human plasma samples from early Alzheimer's disease patients and healthy individuals, and the use of sophisticated statistical tools, identified some metabolic pathways and plasma biomarkers [51]. Nina P Paynter's work shows metabolomics also has important applications in cancer. The processes of life accompany metabolism, such as glycolysis, protein synthesis and metabolism. These fundamental features of cellular metabolism are reprogrammed in cancer cells to support their pathological levels of growth and proliferation. Metabolic reprogramming in malignant cells is likely the result of the multifactorial effects of genomic alterations (i.e. mutations of oncogenes and tumor suppressors), the tumor microenvironment (which imposes metabolic stress caused by compromised nutrients and oxygen availability), and other influences [52]. These changes may be the result of changes in the genome or environmental impacts and a variety of other factors. We need to understand the complete breadth of metabolic abnormalities in cancer because some metabolic changes provide opportunities to develop novel therapeutic targets and predictive biomarkers [52]. As mentioned in Yousra Ahmed-Salim's study, generally, combinations of more than one significant metabolite as a panel, in different studies, achieved a higher sensitivity and specificity for diagnosis than a single metabolite [53]. Metabolomics has become the most powerful platform for studying tissue samples. A common application of metabolomics is the discovery of biomarkers for diagnosis or prediction of treatment sensitivity and prognosis. For example, Yousra Ahmed-Salim et al. conducted a systematic review of the application of metabolomics in the treatment of ovarian cancer. The most frequently described metabolite difference between the biological fluids and tissues of patients with ovarian cancer and those of healthy controls have been in phospholipids [53]. Su et al. interrogated metabolomics and gene-expression from the NCI-60 cell lines to study relationships between metabolite and transcripts [54]. They observed that the metabolome can distinguish cancer subtypes and that metabolite levels correlate well with gene expression under strong correlation models [54]. In conclusion, metabolomics can more accurately determine pathophysiological changes of diseases and identify effective biomarkers through the high-throughput study of metabolites in organisms with abundant sources of samples, so as to further understand the molecular mechanism of diseases. Thus, it is beneficial to the prevention, diagnosis and treatment of diseases.

6. Conclusion

Metabolomics as the important aspect of phenomics is emerging as the frontier field in life science and medical science. Many biological samples have been used to measure metabolomic variations, including extracts from different cells, tissues, organisms, and body-fluids (for example, urine, serum/plasma, tear, exhaled air, saliva, synovial fluid, CSF, and sputum). Metabolomics is classified into targeted metabolomics and untargeted metabolomics. Targeted metabolomics is used to analyze the known metabolite profiling with SRM/MRM methods. Untargeted metabolomics is used to globally analyze the unknown metabolite profiling with NMR-based methods (1D-NMR, 2D-NMR, and 3D-NMR) and MS-based methods (DI-MS, LC–MS, GC–MS, CE-MS, and IM-MS). Metabolomics has been extensively applied in the research and practice of life science and medical science. However, currently the studies on metabolomic variations are much insufficient in the width and depth. The development of high-sensitivity, high-throughput, and high-reproducibility methodology is needed to maximize the coverage of metabolomic variations for clarification of molecular mechanism of a disease, determination of effective therapeutic targets, and discovery of reliable biomarkers for prediction, diagnosis, and prognostic assessment in the context of PPPM practice.

Acknowledgements

The authors acknowledge the financial supports from the Shandong First Medical University Talent Introduction Funds (to X.Z.), the Hunan Provincial Hundred Talent Plan (to X.Z.), and the Hunan Provincial High-Level Health Talents "225" Plan – Medical Academic Leader Funds (to X.Z.).

Author's contributions

X.Z. conceived the concept, designed the manuscript, wrote and critically revised the manuscript, coordinated and was responsible for the correspondence work and financial support. J.Y., S.Z., N.L., and N.L. participated in the literature analysis, and wrote partial manuscript.

Conflict of interest

We declare that we have no financial and personal relationships with other people or organizations.

Acronyms and abbreviations

CE-MS	Capillary electrophoresis coupled with mass spectrometry
CSF	Cerebrospinal fluid
DI-MS	Direct injection coupled with mass spectrometry

GC–MS	Gas chromatography coupled with mass spectrometry
IM-MS	Ion mobility coupled with mass spectrometry
LC–MS	liquid chromatography coupled with mass spectrometry
LC	Liquid chromatography
MS	Mass spectrometry
m/z	Mass-to-charge
NMR	Nuclear magnetic resonance
QqQ-MS	Triple quadrupole mass spectrometry
PPPM	Predictive, preventive and personalized medicine
PTM	Posttranslational modification
SRM/MRM	Selected/multiple reaction monitoring
1D-NMR	One-dimensional NMR
2D-NMR	Two-dimensional NMR
3D-NMR	Three-dimensional NMR
3P medicine/PPPM	Predictive, preventive and personalized medicine

Author details

Xianquan Zhan[1,2]*, Jingru Yang[2], Shu Zheng[2], Nannan Li[1,2] and Na Li[2]

1 Shandong Key Laboratory of Radiation Oncology, Cancer Hospital of Shandong First Medical University, Jinan, Shandong, P.R. China

2 Medical Science and Technology Innovation Center, Shandong First Medical University, Jinan, Shandong, P.R. China

*Address all correspondence to: yjzhan2011@gmail.com

References

[1] Zhan X, Long Y, Lu M. Exploration of variations in proteome and metabolome for predictive diagnostics and personalized treatment algorithms: Innovative approach and examples for potential clinical application. J Proteomics, 2018, 188: 30-40. DOI: 10.1016/j.jprot.2017.08.020.

[2] Li N, Desiderio DM, Zhan X. The use of mass spectrometry in a proteome-centered multiomics study of human pituitary adenomas. Mass Spectrom Rev, 2021, 40: 1-50. DOI:10.1002/mas.21710.

[3] Zhan X (ed.). Proteoforms: Concept and Applications in Medical Sciences. InTech - Open science publisher, London, United Kingdom. Published: July 15th 2020. ISBN: 978-1-83880-034-5. Copyright year: 2020. DOI: 10.5772/intechopen.83687.

[4] Nicholson JK, Lindon JC, Holmes E. "Metabonomics": understanding the metabolic responses of living systems to pathophysiological stimuli via multivariate statistical analysis of biological NMR spectroscopic data. Xenobiotica, 1999, 29: 1181-1189. DOI: 10.1080/004982599238047.

[5] Dunn WB, Bradhurst DI, Atherton HJ, Goodacre R, Griffin JL. Systems level studies of mammalian metabolomes: the roles of mass spectrometry and nuclear magnetic resonance spectroscopy. Chem Soc Rev, 2011, 40: 387-426. DOI: 10.1039/b906712b

[6] Khamis MM, Adamko DJ, El-Aneed A. Mass spectrometric based approaches in urine metabolomics and biomarker discovery. Mass Spectrom Rev, 2017, 36: 115-134. DOI: 10.1002/mas.21455

[7] Tebani A, Abily-Donval L, Afonso C, Marret S, Bekri S. Clinical metabolomics: the new metabolic window for inborn errors of metabolism investigations in the post-genomic era. Int J Mol Sci, 2016, 17: 1167. DOI: 10.3390/ijms17071167

[8] Everett JR. Pharmacometabonomics in humans: a new tool for personalized medicine. Pharmacogenomics, 2015, 16: 737-754. DOI: 10.2217/pgs.15.20

[9] Pareek V, Tian H, Winograd N, Benkovic SJ. Metabolomics and mass spectrometry imaging reveal channeled de novo purine synthesis in cells. Science, 2020, 368(6488): 283-290. doi: 10.1126/science.aaz6465.

[10] Cao G, Song Z, Hong Y, Yang Z, Song Y, Chen Z, Chen Z, Cai Z. Large-scale targeted metabolomics method for metabolite profiling of human samples. Anal Chim Acta, 2020, 1125: 144-151. doi: 10.1016/j.aca.2020.05.053.

[11] Delgado-Povedano MM, Castillo-Peinado LS, Calderón-Santiago M, Luque de Castro MD, Priego-Capote F. Dry sweat as sample for metabolomics analysis. Talanta, 2020, 208: 120428. doi: 10.1016/j.talanta.2019.120428.

[12] Fernández-Ochoa Á, Borrás-Linares I, Quirantes-Piné R, Alarcón-Riquelme ME, Beretta L, Segura-Carretero A; Precisesads Clinical Consortium. Discovering new metabolite alterations in primary sjögren's syndrome in urinary and plasma samples using an HPLC-ESI-QTOF-MS methodology. J Pharm Biomed Anal, 2020, 179: 112999. doi: 10.1016/j.jpba.2019.112999.

[13] Brown AL, Sok P, Taylor O, Woodhouse JP, Bernhardt MB, Raghubar KP, Kahalley LS, Lupo PJ, Hockenberry MJ, Scheurer ME. Cerebrospinal Fluid Metabolomic Profiles Associated With Fatigue During Treatment for Pediatric Acute

Lymphoblastic Leukemia. J Pain Symptom Manage, 2021, 61(3): 464-473. doi: 10.1016/j.jpainsymman.2020.08.030.

[14] Turunen S, Puurunen J, Auriola S, Kullaa AM, Kärkkäinen O, Lohi H, Hanhineva K. Metabolome of canine and human saliva: a non-targeted metabolomics study. Metabolomics, 2020, 16(9): 90. doi: 10.1007/s11306-020-01711-0.

[15] Zhu T, Li S, Wang J, Liu C, Gao L, Zeng Y, Mao R, Cui B, Ji H, Chen Z. Induced sputum metabolomic profiles and oxidative stress are associated with chronic obstructive pulmonary disease (COPD) severity: potential use for predictive, preventive, and personalized medicine. EPMA J, 2020, 11(4): 645-659. doi: 10.1007/s13167-020-00227-w.

[16] Nagana Gowda GA, Raftery D. Analysis of Plasma, Serum, and Whole Blood Metabolites Using 1H NMR Spectroscopy. Methods Mol Biol, 2019, 2037: 17-34. doi: 10.1007/978-1-4939-9690-2_2.

[17] Want EJ, Wilson ID, Gika H, Theodoridis G, Plumb RS, Shockcor J, Holmes E, Nicholson JK. Global metabolic profiling procedures for urine using UPLC-MS. Nat Protoc, 2010, 5: 1005-1018. DOI: 10.1038/nprot.2010.50

[18] Dunn WB, Broadhurst D, Begley P, Zelena E, Francis-McIntyre S, Anderson N, Brown M, Knowles JD, Halsall A, Haselden JN, Nicholls AW, Wilson ID, Kell DB, Goodacre R. Human Serum Metabolome (HUSERMET) Consortium: procedures for large-scale metabolic profiling of serum and plasma using gas chromatography and liquid chromatography coupled to mass spectrometry. Nat Protoc, 2011, 6: 1060-1083. DOI: 10.1038/nprot.2011.335

[19] Ahamad SR, Raish M, Yaqoob SH, Khan A, Shakeel F. Metabolomics and trace element analysis of camel tear by GC-MS and ICP-MS. Biol Trace Elem Res, 2017, 177(2): 251-257. DOI: 10.1007/s12011-016-0889-7

[20] Beale DJ, Jones OA, Karpe AV, Dayalan S, Oh DY, Kouremenos KA, Ahmed W, Palombo EA. A review of analytical techniques and their application in disease diagnosis in breathomics and salivaomics research. Int J Mol Sci, 2016, 18(1): 24. DOI: 10.3390/ijms18010024

[21] Boots AW, Bos LD, van der Schee MP, van Schooten FJ, Sterk PJ. Exhaled molecular fingerprinting in diagnosis and monitoring: validating volatile promises. Trends Mol Med, 2015, 21: 633-644. DOI: 10.1016/j.molmed.2015.08.001

[22] Mikkonen JJ, Singh SP, Herrala M, Lappalainen R, Myllymaa S, Kullaa AM. Salivary metabolomics in the diagnosis of oral cancer and periodontal diseases. J Periodontal Res, 2016, 51: 431-437. DOI: 10.1111/jre.12327

[23] Mickiewicz B, Kelly JJ, Ludwig TE, Weljie AM, Wiley JP, Schmidt TA, Vogel HJ. Metabolic analysis of knee synovial fluid as a potential diagnostic approach for osteoarthritis. J Orthop Res, 2015, 33: 1631-1638. DOI: 10.1002/jor.22949

[24] Graham SF, Chevallier OP, Roberts D, Hölscher C, Elliott CT, Green BD. Investigation of the human brain metabolome to identify potential markers for early diagnosis and therapeutic targets of Alzheimer's disease. Anal Chem, 2013, 85: 1803-1811. DOI: 10.1021/ac303163f

[25] Siskos AP, Jain P, Römisch-Margl W, Bennett M, Achaintre D, Asad Y, Marney L, Richardson L, Koulman A, Griffin JL, Raynaud F, Scalbert A, Adamski J, Prehn C, Keun HC. Interlaboratory reproducibility of a targeted metabolomics platform for

analysis of human serum and plasma. Anal Chem, 2017, 89: 656-665. DOI: 10.1021/acs.analchem.6b02930

[26] Mizuno H, Ueda K, Kobayashi Y, Tsuyama N, Todoroki K, Min JZ, Toyo'oka T. The great importance of normalization of LC-MS data for highly-accurate nontargeted metabolomics. Biomed Chromatogr, 2017, 31: 1. DOI: 10.1002/bmc.3864.DOI: 10.1002/bmc.3864

[27] Kitteringham NR, Jenkins RE, Lane CS, Elliott VL, Park BK. Multiple reaction monitoring for quantitative biomarker analysis in proteomics and metabolomics. J Chromatogr B Anal Technol Biomed Life Sci, 2009, 877: 1229-1239. DOI: 10.1016/j.jchromb.2008.11.013

[28] Zhou J, Yin Y. Strategies for large-scale targeted metabolomics quantification by liquid chromatography-mass spectrometry. Analyst, 2016, 141: 6362-6373. DOI: 10.1039/c6an01753c

[29] Naz S, Moreira dos Santos DC, García A, Barbas C. Analytical protocols based on LC-MS, GC-MS and CE-MS for nontargeted metabolomics of biological tissues. Bioanalysis, 2014, 6: 1657-1677. DOI: 10.4155/bio.14.119

[30] Marchand J, Martineau E, Guitton Y, Dervilly-Pinel G, Giraudeau P. Multidimensional NMR approaches towards highly resolved, sensitive and highthroughput quantitative metabolomics. Curr Opin Biotechnol, 2016, 43: 49-55. DOI: 10.1002/bmc.3864

[31] Kruk J, Doskocz M, Jodłowska E, Zacharzewska A, Łakomiec J, Czaja K, Kujawski J. NMR techniques in metabolomic studies: a quick overview on examples of utilization. Appl Magn Reson, 2017, 48: 1-21. DOI: 10.1007/s00723-016-0846-9

[32] Týčová A, Ledvina V, Klepárník K. Recent advances in CE-MS coupling: instrumentation, methodology, and applications. Electrophoresis, 2017, 38: 115-134. DOI: 10.1002/elps.201600366

[33] Ramautar R, Somsen GW, de Jong GJ. CE-MS for metabolomics: developments and applications in the period 2014-2016. Electrophoresis, 2017, 38: 190-202. DOI: 10.1002/elps.201600370

[34] Mastrangelo A, Ferrarini A, Rey-Stolle F, García A, Barbas C. From sample treatment to biomarker discovery: a tutorial for untargeted metabolomics based on GC-(EI)-Q-MS. Anal Chim Acta, 2015, 900: 21-35. DOI: 10.1016/j.aca.2015.10.001

[35] Kohler I, Giera M. Recent advances in liquid-phase separations for clinical metabolomics. J Sep Sci, 2017, 40: 93-108. DOI: 10.1002/jssc.201600981

[36] Markley JL, Brüschweiler R, Edison AS, Eghbalnia HR, Powers R, Raftery D, Wishart DS. The future of NMR-based metabolomics. Curr Opin Biotechnol, 2016, 43: 34-40. DOI: 10.1016/j.copbio.2016.08.001

[37] Aretz I, Meierhofer D. Advantages and pitfalls of mass spectrometry based metabolome profiling in systems biology. Int J Mol Sci, 2016, 17(5): 632. DOI: 10.3390/ijms17050632

[38] Uppal K, Walker DI, Liu K, Li S, Go YM, Jones DP. Computational metabolomics: a framework for the million metabolome. Chem Res Toxicol, 2016, 29: 1956-1975. DOI: 10.1021/acs.chemrestox.6b00179

[39] Zampieri M, Sekar K, Zamboni N, Sauer U. Frontiers of high-throughput metabolomics. Curr Opin Chem Biol, 2017, 36: 15-23. DOI: 10.1016/j.cbpa.2016.12.006

[40] Fiehn O. Metabolomics – the link between genotypes and phenotypes. Plant Mol Biol, 2002, 48: 155-171.

[41] Lodge S, Nitschke P, Kimhofer T, Coudert JD, Begum S, Bong SH, Richards T, Edgar D, Raby E, Spraul M, Schaefer H, Lindon JC, Loo RL, Holmes E, Nicholson JK. NMR Spectroscopic Windows on the Systemic Effects of SARS-CoV-2 Infection on Plasma Lipoproteins and Metabolites in Relation to Circulating Cytokines. J Proteome Res, 2021, 20(2):1382-1396. doi: 10.1021/acs.jproteome.0c00876.

[42] Kim HC, Baek KH, Ko YJ, Lee HJ, Yim DG, Jo C. Characteristic Metabolic Changes of the Crust from Dry-Aged Beef Using 2D NMR Spectroscopy. Molecules, 2020, 25(13): 3087. doi: 10.3390/molecules25133087.

[43] Wang C, He L, Li DW, Bruschweiler-Li L, Marshall AG, Brüschweiler R. Accurate Identification of Unknown and Known Metabolic Mixture Components by Combining 3D NMR with Fourier Transform Ion Cyclotron Resonance Tandem Mass Spectrometry. J Proteome Res, 2017, 16(10): 3774-3786. doi: 10.1021/acs.jproteome.7b00457.

[44] Hennig K, Abi-Ghanem J, Bunescu A, Meniche X, Biliaut E, Ouattara AD, Lewis MD, Kelly JM, Braillard S, Courtemanche G, Chatelain E, Béquet F. Metabolomics, lipidomics and proteomics profiling of myoblasts infected with Trypanosoma cruzi after treatment with different drugs against Chagas disease. Metabolomics, 2019, 15(9): 117. doi: 10.1007/s11306-019-1583-5.

[45] Depke T, Thöming JG, Kordes A, Häussler S, Brönstrup M. Untargeted LC-MS Metabolomics Differentiates Between Virulent and Avirulent Clinical Strains of Pseudomonas aeruginosa. Biomolecules. 2020, 10(7): 1041. doi: 10.3390/biom10071041.

[46] Yan SC, Chen ZF, Zhang H, Chen Y, Qi Z, Liu G, Cai Z. Evaluation and optimization of sample pretreatment for GC/MS-based metabolomics in embryonic zebrafish. Talanta. 2020 Jan 15;207:120260. doi: 10.1016/j.talanta.2019.120260. Epub 2019 Aug 14. PMID: 31594598.

[47] Segers K, Zhang W, Aourz N, Bongaerts J, Declerck S, Mangelings D, Hankemeier T, De Bundel D, Vander Heyden Y, Smolders I, Ramautar R, Van Eeckhaut A. CE-MS metabolic profiling of volume-restricted plasma samples from an acute mouse model for epileptic seizures to discover potentially involved metabolomic features. Talanta, 2020, 217: 121107. doi: 10.1016/j.talanta.2020.121107.

[48] King AM, Trengove RD, Mullin LG, Rainville PD, Isaac G, Plumb RS, Gethings LA, Wilson ID. Rapid profiling method for the analysis of lipids in human plasma using ion mobility enabled-reversed phase-ultra high performance liquid chromatography/mass spectrometry. J Chromatogr A, 2020, 1611: 460597. doi: 10.1016/j.chroma.2019.460597.

[49] Suhre K, Meisinger C, Döring A, Altmaier E, Belcredi P, Gieger C, Illig T. Metabolic footprint of diabetes: A multiplatform metabolomics study in an epidemiological setting. PloS One, 2010, 5(11): e13953. doi:10.1371/journal.pone.0013953.

[50] Toledo JB, Arnold M, Kastenmüller G, Chang R, Baillie RA, Han X, Kaddurah-Daouk R. Metabolic network failures in Alzheimer's disease: A biochemical road map. Alzheimer's & Dementia, 2017, 13(9): 965-984. doi:10.1016/j.jalz.2017.01.020.

[51] Peña-Bautista C, Roca M, Hervás D, Cuevas A, López-Cuevas R, Vento M, Baquero M, García-Blanco A, Cháfer-Pericás C. Plasma metabolomics

in early Alzheimer's disease patients diagnosed with amyloid biomarker. J Proteomics, 2019, 200: 144-152. doi: 10.1016/j.jprot.2019.04.008.

[52] Kaushik AK, DeBerardinis RJ. Applications of metabolomics to study cancer metabolism. Biochim Biophys Acta Rev Cancer, 2018, 1870(1): 2-14. doi: 10.1016/j.bbcan.2018.04.009.

[53] Ahmed-Salim Y, Galazis N, Bracewell-Milnes T, Phelps DL, Jones BP, Chan M, Munoz-Gonzales MD, Matsuzono T, Smith JR, Yazbek J, Krell J, Ghaem-Maghami S, Saso S. The application of metabolomics in ovarian cancer management: a systematic review. Int J Gynecol Cancer, 2021, 31(5): 754-774. doi: 10.1136/ijgc-2020-001862.

[54] Su G, Burant CF, Beecher CW, Athey BD, Meng F. Integrated metabolome and transcriptome analysis of the NCI60 dataset. BMC Bioinformatics, 2011, 12 Suppl 1 (Suppl 1): S36. doi: 10.1186/1471-2105-12-S1-S36.

Chapter 2

From Targeted Quantification to Untargeted Metabolomics

Veronica Lelli, Antonio Belardo and Anna Maria Timperio

Abstract

Metabolomics is an emerging and rapidly evolving technology tool, which involves quantitative and qualitative metabolite assessments science. It offers tremendous promise for different applications in various fields such as medical, environmental, nutrition, and agricultural sciences. Metabolomic approach is based on global identification of a high number of metabolites present in a biological fluid. This allows to characterize the metabolic profile of a given condition and to identify which metabolites or metabolite patterns may be useful in the discrimination between different groups. The use of one mass spectrometry (MS) platform from targeted quantification to untargeted metabolomics will make more efficient workflows in many fields and should allow projects to be more easily undertaken and realized. Metabolomics can be divided into *non-targeted* and *targeted*. The first one can analyze metabolites derived from the organisms comprehensively and systematically, so it is an unbiased metabolomics analysis that can discover new biomarkers. Targeted metabolomics, on the other hand, is the study and analysis of specific metabolites. Both have their own advantages and disadvantages, and are often used in combination for discovery and accurate weight determination of differential metabolites, and allow in-depth research and analysis of subsequent metabolic molecular markers. Targeted and non-targeted metabolomics are involved in food identification, disease research, animal model verification, biomarker discovery, disease diagnosis, drug development, drug screening, drug evaluation, clinical plant metabolism and microbial metabolism research. The aim of this chapter is to highlight the versatility of metabolomic analysis due to both the enormous variety of samples and the no strict barriers between quantitative and qualitative analysis. For this purpose, two examples from our group will be considered. Using non-targeted metabolomics in opposite Antarctic *cryptoendolytic* communities exposed to the sun, we revealed specific adaptations. Instead, through the targeted metabolomics applied to the urine during childbirth, we identified a different distribution of specific metabolites and the metabolic differences allowed us to discriminate between the two phases of labor, highlighting the metabolites most involved in the discrimination. The choice of these two approaches is to highlight that metabolomic analysis can be applied to any sample, even physiologically and metabolomically very distant, as can be microorganisms living on Antarctic rocks and biological fluids such as urine.

Keywords: urine, metabolite, HPLC–MS, fungus, biomarkers, system biology

1. Introduction

1.1 Metabolomics: the medicine of the third millennium

Metabolomics is a discipline thanks to whose analysis it is possible to ascertain the presence of biochemical imbalances caused by the lack of nutrients that are the basis of the functions of our body. With this research it is therefore possible to identify the true causes of any chronic disease and restore the biochemical balance of our body.

The use of "omics" sciences, especially metabolomics, has been having positive implications in recent years in the main actions of daily life, as monitoring the metabolism helps to keep energy levels, sleep and body weight under control. The operation is very simple: by measuring the metabolites present in the body, problems are identified and action is taken in a targeted and relevant way. An action of extraordinary effectiveness if you think about the current situation, for example in the food sector; nutrition today is very rich in calories but poor in nutrients, with the real risk of being overfed, but undernourished (**Figure 1**).

It is essential to remember that this is not an alternative medicine, but is complementary to other disciplines: in the face of even important pathologies, such as neoplasms, it will support the oncologist, improving the responses to cancer treatment and helping to defeat the disease without interfering with the pathways of oncological treatment. A normal blood draw or a simple urine sample is sufficient to check the metabolites. An extraordinarily precise picture of the situation will emerge, a sort of fingerprint of our body and of how many external factors,

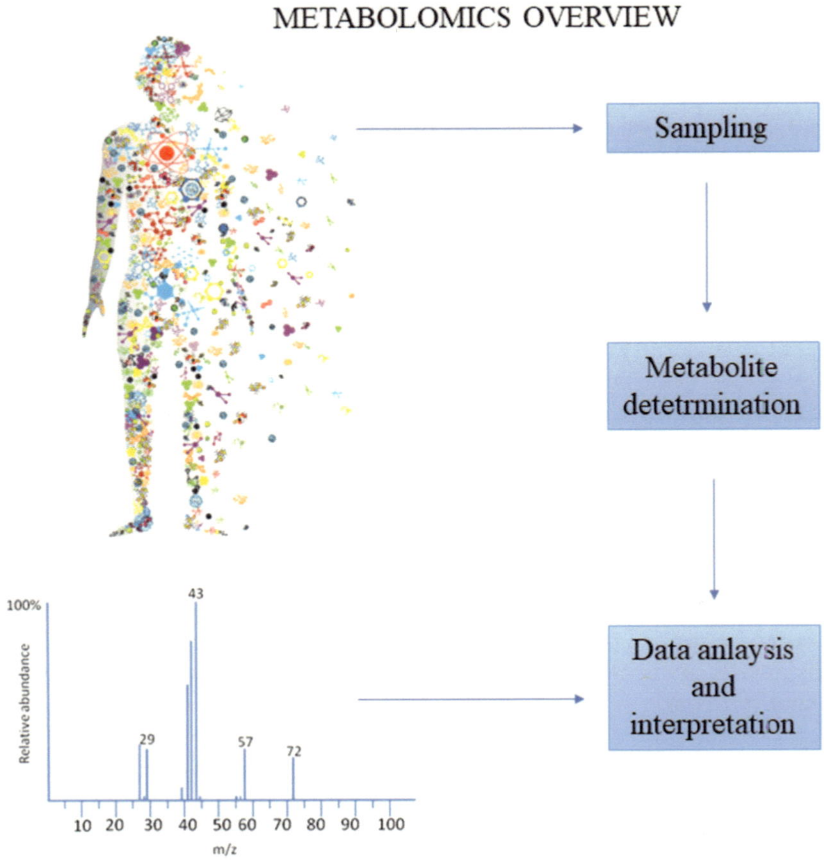

Figure 1.
Metabolomic workflow.

which we can define as part of the so-called epigenetics, can directly influence the chemical aspect. The ultimate goal is to bring the body back to perfect efficiency, taking into account that the body tends by itself, by its nature, to the best possible state of health. However, when the body has difficulty repairing the damage or is facing a progressive problem, which does not have time to fix or lacks adequate resources, metabolomics comes to its rescue. Our body must be considered like a car: to function at its best, the best fuel is needed together with winning strategies of suitable equipment. This discipline, which is progressively affirming itself, has set itself the task of identifying the optimal conditions to support the architrave of human existence.

2. Targeted and untargeted metabolomics

The metabolome is the final downstream product of the genome and consists of all low molecular weight molecules (metabolites) in a cell, tissue or organism [1, 2]. The metabolic profile can provide a complete picture of that cell's physiology. As emerging data suggest an important role for the microbiome and its metabolic products, the potential size of the metabolome is often highly controversial. Given the variety of chemical classes and physical properties that characterize metabolites and the dynamic range of metabolite concentrations over large orders of magnitude, a wide range of analytical techniques are required for metabolomics research. Metabolomics aims to identify and quantify multiple molecules in the context of physiological stimuli or in disease states. The "omics" revolution of the 1980s and 1990s provided new methodologies for the study of interactions on a global level and offered an alternative means of investigation to the more reductionist one in molecular biology. Omics is a field that aims to study the abundance and/or structural characterization of a wide range of molecules in organisms in distinct scenarios. In the clinical field, high-throughput omics techniques are used for disease characterization to better predict the clinical course of organisms and to evaluate the efficacy of existing or developing therapies [3]. In food science, for example, omics plays a significant role in trying to improve human nutrition [4]. On the other hand, concerning the environment, omics studies aim to evaluate the alterations that organisms could undergo after exposure to environmental stressors [5, 6]. In recent years, a variety of omics subdisciplines have emerged (eg Fluxomics, lipidomics, glycomics, foodomics, interactomics and metallomics), demonstrating that omics is a continuously evolving discipline and among all these platforms, metabolomics is becoming increasingly popular [7].

The first definition of metabolomics dates back to the 1990s, describing techniques aimed at identifying existing metabolites within a cell, tissue or organism during a genetic alteration or physiological stimulus [8, 9]. Metabolomics has been shown to be complementary to other omics techniques, thus identifying - called silent phenotypes - genes that when perturbed have no apparent influence on physical characteristics or behavior [10]. The metabolomic approach can be conducted in two distinct ways; non-targeted approach and targeted approach [11].

The reason for this differentiation is due to the different types of data generated in these two approaches, which must be handled accordingly (**Figure 2**). Targeted studies focus research on a number of known metabolites, while non-targeted studies allow for a more comprehensive evaluation of metabolomic profiles. Most of the methodologies used in the first targeted studies only allowed for the identification of a limited number of metabolites. However, recent targeted methodologies allow for the creation of large-scale metabolic profiles, including hundreds of compounds. However, the number of compounds analyzed in non-targeted studies is even greater. This is because entire datasets, including thousands of metabolic signals, need to be processed, and of these, few are finally identified as candidate biomarkers.

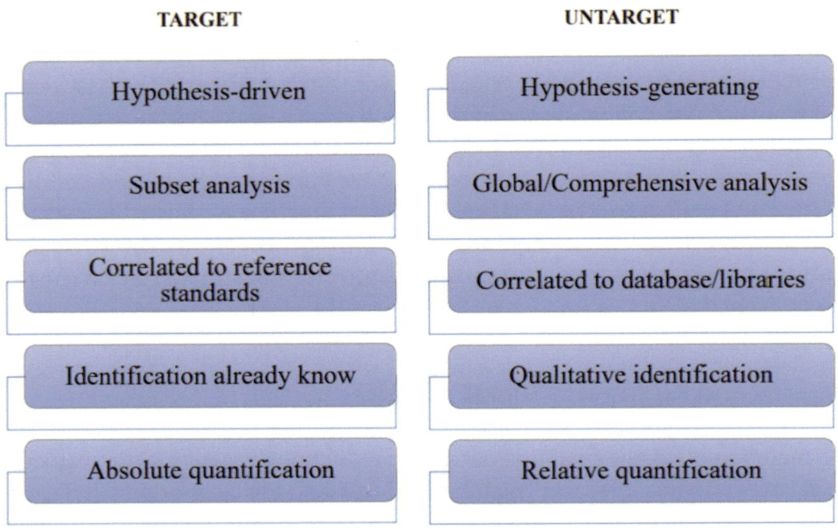

Figure 2.
Targeted and untargeted approach.

3. Untargeted approaches

Untargeted approaches provide the most correct path to detect unexpected changes in metabolite concentrations. The goal is to maximize the number of metabolites detected and thus provide the opportunity to observe unexpected changes. However, a single analytical method cannot detect all metabolites in a biological system. It is therefore necessary to combine multiple analytical approaches (such as complementary HPLC methods) to maximize the number of metabolites detected and improve metabolome coverage.

Sample preparation in non-targeted studies consists of extracting the metabolites from the biological sample in a suitable solvent for analytical analysis. The extracted sample is analyzed with an appropriate analytical method (for example, LC–MS). The result of the mass spectrometry analysis is a chromatogram and the peak area of each metabolite is used as a parameter in the statistical analysis to define the concentration differences between the different biological samples measured. This is called relative quantification as there is no comparison with calibration curves constructed from chemical standards. The use of calibration curves is indispensable for full quantification. The biological significance of each metabolite is determined during data analysis and metabolite identification, and biological interpretation is performed at the end of the experimental pipeline. Currently, one of the main limitations in non-targeted approaches is the identification of metabolites. It may not be possible to identify the metabolites highlighted in the statistical analysis as significant changes between biological classes in the study. The identification of metabolites is currently one of the hot topics of metabolomics.

4. Targeted approaches

Targeted studies investigate a relatively limited and specific number of metabolites. At the start of the study, before data acquisition is performed, the metabolites are chemically characterized and biochemically annotated. Targeted methods have greater selectivity and sensitivity than non-targeted methods. A targeted study can only be performed if a genuine chemical standard of the metabolite is available. The quantification of metabolites is performed using internal and chemical standards to construct calibration curves for each of the metabolites under study. Sample preparation in targeted studies applies methods that can be optimized to retain metabolites of interest and to remove other biological species and analytical artifacts that are not performed via downstream analysis.

Therefore, data analysis strategies for non-targeted studies require very extensive chromatogram processing. A large number of data analysis strategies are found in the literature, but none of them can be considered the optimal choice in all cases, which makes data analysis an open task in bioinformatics research. In fact, the field of MS-based metabolomics is quite young and new methods, software and platforms are regularly published or updated.

The data are produced by chemical–physical investigation techniques such as magnetic resonance spectroscopy, chromatography and mass spectrometry applied to biofluid samples or suitably selected solid tissues. These methodologies find today numerous possibilities of use in the field of medical sciences where there are numerous variables detectable on human and animal subjects that present a specific pathology. This approach is valid both for the description of existing pathologies and for the identification of pre-pathological stages. The use of metabolomic methods can help provide an overall - holistic - view of the problem, highlighting the relationships between variables and their relative importance, and can also highlight differences and similarities between samples. Considering individual biological processes as isolated processes expresses a reductionist view of vital functions, an abstraction that at times makes it possible to considerably simplify the problem under consideration but which inevitably leads to models of limited value.

5. Metabolomics techniques

There is a range of analytical chemistry tools applied in metabolomics research.

Each instrument has advantages and limitations, and no single instrument or instrumental method can detect all of the metabolites present in a metabolome, but multiple instrumental methods or multiple different instruments are required to provide the largest number of metabolites detected.

Several analytical techniques have been developed for each of the omics platforms, including techniques based on DNA microarray and RNA sequencing [12], nuclear magnetic resonance (NMR) spectroscopy [13, 14] and mass spectrometry (MS) [15, 16]. NMR and MS are the most used in the field of metabolomics. High resolution proton NMR spectroscopy (1H-NMR) has proven to be one of the best technologies for examining biofluids and studying intact tissues, as the result is a complete signal profile of metabolites without separation, derivatization and pre-selected measurement parameters [17, 18]. On the other hand, MS methods, both by direct injection [19] and coupled with chromatographic techniques [20], have also evolved into an excellent technology for metabolomics due to their ability to analyze low molecular weight compounds in biological systems. These two approaches (NMR and MS) are complementary and the integration of both technologies can provide more comprehensive information in the field of metabolomics.

Applications of metabolomics have expanded in line with genomics, proteomics and transcriptomics with the aim of determining gene function in microbes [21], plants [22] and animals [23]. Many different applications are also used today, such as the determination of metabolic biomarkers that change as an indicator of the presence of a disease or in response to a pharmacological intervention or the determination of the effect of biochemical or environmental stress on plants or microbes, genetically modified plants [24], bacterial characterization [25], health status assessments [26] and metabolic engineering [27].

5.1 Gas chromatography–mass spectrometry (GC–MS)

GC–MS is a combined system in which volatile and thermally stable compounds are first separated from the GC and then the eluted compounds are detected by electron impact mass spectrometers. During the run, aliquots of derivatized samples (injection volumes of 1 μl or less) are analyzed using split and splitless techniques on different polarity GC columns (DB-5 or DB-50 or similar capillary columns in stationary phase are most commonly used). These provide both high chromatographic resolution of compounds and high sensitivity (typical limits of detection are concentrations of pmol or nmol). Quantification is provided by external calibration or response ratio (metabolite peak area/internal standard peak area). The coverage of the metabolome is largely characterized by the volatility of the non-derivatized or derivatized sample components. The identification of metabolites is provided by matching the retention time and mass spectrum of the sample peak with those of a pure compound previously tested on the same or a different instrument under identical instrumental conditions [28]. Since the electron impact mass spectrometer provides the standard fragmentation of molecular ions during ionization, structural identification can be performed through the interpretation of fragment ions and fragmentation patterns.

5.2 Liquid chromatography-mass spectrometry (LC–MS)

LC–MS provides separation of metabolites by liquid chromatography followed by electrospray ionization (ESI) or, less typically, atmospheric pressure chemical ionization (APCI) [29]. This technique differs from GC–MS for several reasons (lower assay temperatures and unsolicited sample volatility) and this simplifies sample preparation. In most non-pharmaceutical applications, samples are prepared after intracellular extraction and/or protein precipitation by dilution in an appropriate solvent. The chemistry and HPLC column size used will affect chromatographic resolution and sensitivity. Analytical columns do not provide the chromatographic resolving power to separate these complex samples and run times of 10 minutes followed by chemometric instruments are used to extract the chromatographically unsolved data and classify the differences between the samples [30]. The application of very high-pressure chromatography systems can improve the chromatographic resolution. The most common column chemicals used today are reverse phase C 18 or C 8 columns. However, for polar metabolites injected on these columns, the retention of these metabolites is minimal, thus reducing the volume of interpretive data. To overcome this problem, other chemical columns are needed, such as HILIC [31] and other weak ion exchange chemicals. The sample then, once the chromatographic separation has been carried out, reaches the source of the mass spectrometer. Electrospray instrumentation operates in positive and negative ion modes (as separate experiments or by polarity switching during analyses) and detects only those metabolites that can be ionized by adding or removing a proton or adding another species of ions. Metabolites are generally detected in one but not both ion modalities, so broader metabolic coverage can be achieved by analysis in both modalities. Quantification is performed by external calibration. ESI does not cause molecular ion fragmentation as observed in electron impact mass spectrometers, thus it does not allow direct identification of metabolites by comparing ESI mass spectra, as ESI mass spectral libraries are not commonly available (as in the case by GC–MS). However, with the use of accurate mass measurements and/or tandem MS (MS/MS) to provide collision induced dissociation (CID) and correlated mass spectra (MS/MS), correct identification of metabolites can be achieved [30].

5.3 Direct injection mass spectrometry (DIMS)

DIMS is shown as a high-throughput screening tool (hundreds of samples per day with a run time generally of 1 minute). The extracts of the crude sample are injected or infused into an electrospray mass spectrometer and results in a mass spectrum per sample, which is representative of the composition of the sample. The ionization capacity of the metabolite determines metabolic coverage, as for LC–MS. The applications of DIMS are mainly concentrated in the microbial [32] and plant areas.

5.4 Fourier transform infrared spectroscopy (FT-IR)

Vibrational spectroscopy techniques such as Fourier transform infrared spectroscopy (FT-IR) and Raman spectroscopy have been used to analyze metabolic changes in biological samples. The methodologies consist of passing ultraviolet or infrared light through a sample before it is detected. The vibrations and rotations of the bonds relating to different chemical groups resulting from the interaction of the sample with ultraviolet or infrared light are mainly measured.

The problem with these techniques is the inability to detect each metabolite separately; instead, it is a specific technique for single molecules that will absorb ultraviolet or infrared light at specific wavelengths. In FT-IR, a metabolic fingerprint is taken with a single absorption spectrum collected for each sample and consisting of information for many metabolites. The result is similar to data produced by direct infusion mass spectrometry where a single mass spectrum is collected rather than an absorption spectrum for each sample. The metabolic fingerprints produced in these approaches lack the sensitivity of mass spectrometry but are a useful tool for high throughput screening since the FT-IR analysis time is approximately one minute per sample. To date, the vast majority of metabolomic studies undertaken using vibrational spectroscopy have been performed with FT-IR spectroscopy. However, the work was done using Raman and, in terms of metabolomics, this is an emerging technology with significant potential for metabolite monitoring [33].

5.5 Nuclear magnetic resonance (NMR)

Nuclear magnetic resonance (NMR) spectroscopy applies the magnetic properties of atomic nuclei in a metabolite. Only some atoms are active NMRs and include 1H, 13C and 31P. Proton (1H) NMR spectroscopy is the most frequently applied in metabolomics. The technique works by inserting a liquid sample into a small internal diameter tube (about 5 mm), where it is pulsed with a range of radio frequencies that cover all possible energies required to excite the selected type of nuclei. Nuclei absorb energy at different radio frequencies depending on their chemical environment and then the release of this energy is measured, forming what is called a free induction decay (FID). This FID is converted from a time domain data set to a frequency domain - using a Fourier transformation - and an NMR spectrum is constructed as the absorption energy plotted against the peak intensity. The NMR spectrum (in particular the chemical shift) depends on the effect of the shielding by the electrons orbiting the nucleus. The chemical shift for 1 H NMR is determined as the difference (in ppm) between the resonance frequency of the observed proton and that of a reference proton present in a reference compound (for 1 H NMR experiments, tetramethyl-silane in solution, fixed at 0 ppm).

The intensity of the signal depends on the number of identical nuclei and the presence of complex samples does not interfere with the measured intensity, as does the suppression of ionization with electrospray ionization. NMR spectroscopy is a high-speed fingerprinting technique. Raw samples are mixed with a solution of the reference compound added to an NMR probe (usually less than 2 mL), inserted into the instrument and analyzed. NMR probes are generally based on a large volume of μl and this adds constraints to the required sample volume. However, the introduction of 1 mm μl probes allowed to analyze volumes of 2 μl, thus enabling invasive sampling of smaller volumes of study subjects, which is important for small animal studies [10]. Spectra are complex and contain thousands of metabolic signals. For data processing, the spectrum is generally divided into chemical shift ranges with widths of 0.02–0.04 ppm. All signals in this bucket are added together. Chemical changes can be assigned to specific metabolites and the pure metabolite can be added for further clarification. However, the spectrum model is generally used in sample classification, similar to that used for FT-IR and DIMS.

6. Examples of metabolomics studies

6.1 Urine metabolomics

The metabolic profile of biofluids has emerged as an important tool in the diagnosis of numerous diseases that remain silent until late progression of the disease [34]. Because urine is such a rich source of biomarkers, the metabolic profile is a promising tool for assessing therapeutic efficacy. Urine collection is also non-invasive, does not require patient preparation and substantially improves compliance. Recent results clearly demonstrate the potential of urine metabolomics in diagnosis by providing new insights into the biochemistry of its pathophysiology [35]. The qualitative/quantitative analysis on the urine of pregnant women between two different stages of labor called OL (out of labor) and IL-DP (in labor in the dilation phase), using as a technique the ultra-performance liquid chromatography of hydrophilic interaction coupled with mass spectrometry (HILIC-UPLC–MS), a highly sensitive, accurate and unbiased approach, are an example [36]. The list of metabolites is shown in **Table 1**. The urinary metabolites showing the greatest differences belong to the steroid hormone, in particular conjugated estrogens and amino acids, much of this difference being determined by fetal contribution. The increased excretion of conjugated estrogens in the DP stage may confirm the coordinated role played between fetus, mother and placenta during labor. It is reasonable to consider this terminal phase of pregnancy not only as a mechanical event linked to the increase in uterine contractions, but as a more complex process. **Figure 3** shows the major compounds excreted in the urine (such as Estradiol Glucoronide, Estrone 3-Sulphate and Estriol Glucuronate) which are downstream of a more intricate process. These compounds originate from an interconnected metabolism between mother-fetus and placenta involving steroid hormones. As seen in **Figure 3**, the metabolites excreted can come directly from the mother, or with the contribution of both the placenta and the fetus. In the latter case, the excretion of metabolites occurs through the degradation of intermediates, in particular the hormone pregnenolone.

6.2 Antarctic cryptoendolithic communities

Antarctic *cryptoendolite* communities are microbial ecosystems that dominate the biology of most ice-free areas in mainland Antarctica. These are complex and

Compounds	Molecular weight	Percentage
3 hydroxy2-methyl-1H-quinolin-one	175.05	3% - up
19 chloro19-Chloro-3beta-hydroxyandrost-5-en-17-one = dehydroepiandrosterone	365.1695	95% - up
Androst-5-ene-3beta,17beta-diol = androsterone	290.1736	21% - down
Androsterone	290.1736	21% - down
Dehydroepiandrosterone Sulfate	369.0992	92% - down
Dehydroepiandrosterone	288.1873	3% - down
Pregnanediol	321.2145	25% - down
3-Hydroxy-1-methylestra-1,3,5(10),6-tetraen-17-one	283.1551	63% - up
Tetrahydrocortisone	365.1592	88% - down
Estrone 3 sulfate	351.1085	>100% - up
Estrone gluconoride	447.1790	87% - down
Estradiol 17 beta 3 gluconoride	449.1945	42% - up
Ser	105.09	59% - down
Val	117.15	35% - up
His	155.15	23% - up
Arg	174.20	11% - up
Cys	121.16	47% - up
Ala	89.09	47% - down
Glu	147.13	85% - up
Gln	146.14	25% - up
Leu	131.17	38% - up
Lys	146.19	60% - up
Ile	131.17	38% - up
Thr	119.12	18% - up
Phe	165.19	11% - down
Tyr	181.19	12% - down

Table 1.
Estrogen and amino acid amount extracted from the urine in the two stages of pregnancy (OL and IL-DP). The table refers to the relative concentration of the metabolites calculating as a percentage of the total compounds by comparing the intensity of deconvolution of each compound.

self-supporting assemblies formed by the association of autotrophic and heterotrophic microorganisms such as Bacteria Chlorophyta and Fungi, which live at the limit of their physiological adaptability and this represent the only possibility of survival before extinction. They live inside the pores of the rocks creating an environment that protects them from environmental stress, they are extremely tolerant and remarkably resistant to stress and finally they adapt perfectly to the lithic life (**Figure 4A**).

The study was conducted on the basis of the different solar exposure, verifying how this factor influences the production of key metabolites and therefore on the adaptation strategies implemented by these microorganisms to survive in extreme conditions (**Figure 4B**). The result is the presence of 331 altered and differentially expressed metabolites [37]. All intermediates of melanogenesis are found in the highest concentration in south facing rocks (**Table 2**). Organisms have developed

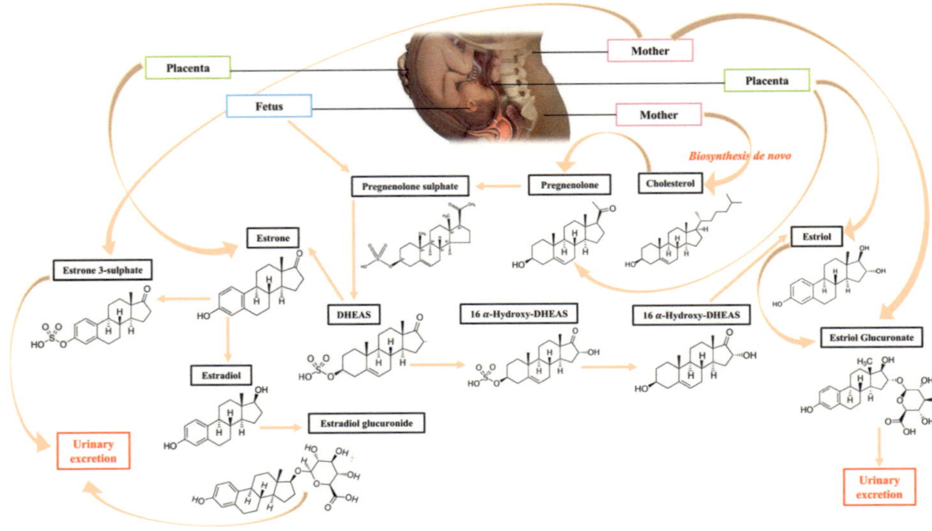

Figure 3.
A coordinated cycle between mother fetus and placenta through the biosynthesis of steroid hormones.

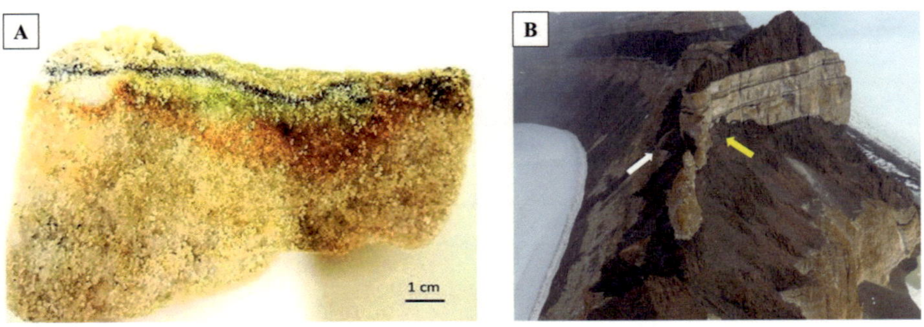

Figure 4.
*A - Communities stratified on rocks. B - Antarctic cryptoendolite communities. The yellow arrow indicates the north-facing surface, while the white arrow indicates the south-facing surface. (**Figure 4B** comes from Coleine et al. [37].*

Compound	Molecular weight	Regulation
Gentisyl alcohol	140.14	Down
Hypoxanthine	136.11	Down
6- Methoxymellein	208.21	Down
Allantoin	158.12	Down
DOPA	197.18	Up
5,6-Hydroxy indole	133.15	Up
Anthocyanin 3-O-β-D glucoside	449.38	Up
Plastoquinone	749.2	Up
Xanthine	152.11	Down
Uric acid	168.11	Down
Tyrosine	181.19	Up

Compound	Molecular weight	Regulation
Dopaquinone	195.17	Up
Cysteinyldopa	316.33	Down
Dopachrome	193.16	Up
5,6-Dihydroxyindole	149.15	Up
Indole 5,6- quinone	147.13	Up

Table 2.
Regulation of metabolites in Antarctic cryptoendolithic *communities exposed to the south in the absence of sun compared to the north in the presence of minimal solar radiation.*

metabolic profiles responding to the condition of deprivation of sunlight, in fact in fungi, melanin performs functions such as photoprotection, energy conversion, protection from thermal stress, metal chelation, resistance to drying and cell strengthening. Therefore, every organism that lives in non-optimal conditions has the ability to modify its biological functions and its structures according to external dynamics. The adaptation analysis allowed to demonstrate how the environment, and in particular the solar exposure, can strongly influence the olic phenotype of these microbial communities, demonstrating how the direct incidence of sunlight in one case, or their absence in the other, it has determined the acquisition of various adaptations by the colonizing species aimed at favoring their growth and guaranteeing their survival in such a hostile environment.

7. Conclusion

Starting from a holistic approach, with a high-throughput non-targeted metabolomics without prior knowledge of the metabolome, it is possible to find biologically relevant metabolites (potential biomarkers), characteristic of targeted metabolomics as shown by the two examples described in this chapter. In the case of urinary metabolites hundreds of metabolites are represented but the major differences between OL and IL-DP belong to the steroid hormone, in particular to conjugated estrogens and amino acids. These compounds could predict the time to delivery in pregnant women.

As for the Antarctic *cryptoendolytic* communities, among the candidates to be considered biomarkers, the precursor metabolites of the melanin and allantoin pathways emerged as these were the most affected by exposure to the sun. These paths can be considered directly involved in the response to environmental pressure. In conclusion, both targeted and non-targeted metabolomics are then used in combination for the detection and accurate weighting of differential metabolites. Furthermore, the types of metabolites that are recovered are influenced by the extraction and the analytical method chosen. In this chapter the main devices used have been summarized, subsequently the results will require computational tools to identify and correlate the metabolites among the samples to examine their connection in the metabolic pathways in relation to the phenotype.

Conflict of interest

The authors declare no conflict of interest.

Abbreviations

CID	Collision induced dissociation
DIMS	Direct Injection Mass Spectrometry
FID	Free induction decay
FT-IR	Fourier Transform Infrared Spectroscopy
GC–MS	Gas Chromatography–Mass Spectrometry
HILIC	Hydrophilic interaction chromatography
HPLC	High performance liquid chromatography
IL-DP	In labor in the dilation phase
LC–MS	Liquid Chromatography-Mass Spectrometry
MS	Mass spectrometry
NMR	Nuclear magnetic resonance
OL	Out of labor

Author details

Veronica Lelli[†], Antonio Belardo[†] and Anna Maria Timperio[*]
Department of Ecological and Biological Sciences, University of Tuscia, Viterbo, Italy

*Address all correspondence to: timperio@unitus.it

† These authors contributed equally to this work.

References

[1] Goodacre, R. Metabolic Profiling: Its role in Biomarker Discovery and Gene Function Analysis. Kluwer Academic Publishers. Springer Scince; 2003. doi: 10.1007/978-1-4615-0333-0

[2] Goodacre R, Vaidyanathan S, Dunn WB, Harrigan GG, Kell DB. Metabolomics by numbers: acquiring and understanding global metabolite data. Trends Biotechnol. 2004; 22:245-252. doi: 10.1016/j.tibtech.2004.03.007

[3] McShane LM, Cavenagh MM, Lively TG, Eberhard DA, Bigbee WL, Williams PM, et al. Criteria for the use of omics-based predictors in clinical trials: explanation and elaboration. BCM Med. 2013;11:(1) 220. doi: 10.1038/nature12564

[4] Capozzi F, Bordoni A. Foodomics. A new comprehensive approach to food and nutrition. Genes Nutr. 2013;8:1-4. doi: 10.1007/s12263-012-0310-x

[5] Bundy G, Davey MP, Viant MR. Environmental metabolomics: a critical review and future perspectives Metabolomics. 2008;5:3-21. doi: 10.1007/s11306-008-0152-0

[6] Viant MR, Sommer U. Mass spectrometry based environmental metabolomics: a primer and review. Metabolomics. 2012;9:144-158. doi: 10.1007/s11306-012-0412-x

[7] Gorrochategui E, Jaumot J, Lacorte S, Tauler R. Data analysis strategies for targeted and untargeted LC-MS metabolomic studies: Overview and workflow. TrAC Trends in Analytical Chemistry. 2016;82:425-442. doi: 10.1016/j.trac.2016.07.004

[8] Oliver SG, Winson MK, Kell DB, Baganz F. Systematic functional analysis of the yeast genome. Trends Biotechnol. 1998;16:373-378. doi: 10.1016/s0167-7799(98)01214-1

[9] Nicholson JK, Lindon JC, Holmes E. `Metabonomics': understanding the metabolic responses of living systems to pathophysiological stimuli via multivariate statistical analysis of biological NMR spectroscopic data. Xenobiotica. 1998;29:1181-1189. doi: 10.1080/004982599238047

[10] Griffin JL, Nicholls AW, Keun HC, Mortishire-Smith RJ, Nicholson JK, Kuehn T. Metabolic profiling of rodent biological fluids via 1H NMR spectroscopy using a 1 mm microlitre probe. Analyst. 2002;127:582-584. doi: 10.1039/b201207c

[11] Roberts LD, Souza AL, Gerszten RE, and Clish CB. Targeted Metabolomics. Curr Protoc Mol Biol. 2012. doi: 10.1002/0471142727.mb3002s98

[12] Campos B, Garcia-Reyero N, Rivetti C, Escalon L, Habib T, Tauler R, et al. Identification of metabolic pathways in Daphnia magna explaining hormetic effects of selective serotonin reuptake inhibitors and 4-nonylphenol using transcriptomic and phenotypic responses Environ. Sci. Technol. 2013;47:9434-9443. doi: 10.1021/es4012299

[13] Kim HK, Choi YH, Verpoorte R. NMR-based plant metabolomics: where do we stand, where do we go? Trends Biotechnol. 2011;29: 267-275. doi: 10.1016/j.tibtech.2011.02.001

[14] Puig-Castellví F, Alfonso I, Piña B, Tauler R. A quantitative 1H NMR approach for evaluating the metabolic response of *Saccharomyces cerevisiae* to mild heat stress Metabolomics. 2015;11:1612-1625. doi: 10.1007/s11306-015-0812-9

[15] Halket JM. Chemical derivatization and mass spectral libraries in metabolic profiling by GC/MS and LC/MS/MS J. Exp. Bot. 2004;56:219-243. doi: 10.1093/jxb/eri069

[16] DettmerK, Aronov PA, Hammock BD. Mass spectrometry-based metabolomics Mass Spectrom. Rev. 2007;26:51-78. doi: 10.1002/mas.20108

[17] Nicholson JK, Wilson ID. High resolution proton magnetic resonance spectroscopy of biological fluids. Prog Nucl Magn Reson Spectrosc. 1989;21:449-501. doi: 10.1016/0079-6565(89)80008-1

[18] Lindon JC, Holmes E, Nicholson JK. So what's the deal with metabonomics? Anal Chem. 2003;75:384A–3391. doi: 10.1021/ac031386+

[19] Weber RJM, Southam AD, Sommer U, Viant MR. Characterization of isotopic abundance measurements in high resolution FT-ICR and Orbitrap mass spectra for improved confidence of metabolite identification Anal. Chem. 2011;83:3737-3743. doi: 10.1021/ac2001803

[20] Wilson ID, Plumb R, Granger J, Major H, Williams R, Lenz EM. HPLC-MS-based methods for the study of metabonomics. J Chromatogr B Analyt Technol Biomed Life Sci. 2005;817:67-76. doi: 10.1016/j.jchromb.2004.07.045

[21] Raamsdonk LM, et al. Co-consumption of sugars or ethanol and glucose in a Saccharomyces cerevisiae strain deleted in the HXK2 gene. Yeast. 2001;18(11):1023-1033. doi: 10.1002/yea.746

[22] Roessner U, Luedemann A, Brust D, Fiehn O, Linke T, Willmitzer L, Fernie AR. Metabolic profiling allows comprehensive phenotyping of genetically or environmentally modified plant systems. Plant Cell. 2001;13:11-29. doi: 10.1105/tpc.13.1.11

[23] Le Gall G, Colquhoun IJ, Davis AL, Collins GJ, Verhoeyen ME. Metabolite profiling of tomato (Lycopersicon esculentum) using 1H NMR spectroscopy as a tool to detect potential unintended effects following a genetic modification. Journal of Agricultural and Food Chemistry. 2003;51(9):2447-2456. doi: 10.1021/jf0259967

[24] Ballereau S. et al. Functional Genomics, Proteomics, Metabolomics and Bioinformatics for Systems Biology. In: Prokop A., Csukás B. (eds) Systems Biology. Springer, Dordrecht. 2013. doi: 10.1007/978-94-007-6803-1_1

[25] Vaidyanathan S, Kell DB, Goodacre R. Flow-injection electrospray ionization mass spectrometry of crude cell extracts for high-throughput bacterial identification. Journal of the American Society for Mass Spectrometry. 200;13(2):118-128. doi: 10.1016/S1044-0305(01)00339-7

[26] Watkins SM, German JB. Toward the implementation of metabolomic assessments of human health and nutrition. Current Opinion in Biotechnology. 2002;13(5):512-516. doi: 10.1016/s0958-1669(02)00363-4

[27] Teusink B, Passarge J, Reijenga CA, Esgalhado E, Van der Weijden CC, Schepper M, Walsh MC, Bakker BM, Van Dam K, Westerhoff HV, and Snoep JL. Can yeast glycolysis be understood in terms of in vitro kinetics of the constituent enzymes? Testing biochemistry. Eur J Biochem. 2000;267(17):5313-5329. doi: 10.1046/j.1432-1327.2000.01527.x

[28] Wagner C, Sefkow M, Kopka J. Construction and application of a mass spectral and retention time index database generated from plant GC/EI-TOF-MS metabolite profiles. Phytochemistry. 2003;62(6):887-900. doi: 10.1016/s0031-9422(02)00703-3

[29] Bakhtiar R, Ramos L, Tse FLS. High-throughput mass spectrometric analysis of xenobiotics in biological fluids. Journal of

Liquid Chromatography and Related Technologies. 2002;25(4):07-540. doi: 10.1081/JLC-120003344

[30] Lenz EM, Bright J, Knight R, Wilson ID, Major H. Cyclosporin A-induced changes in endogenous metabolites in rat urine: A metabonomic investigation using high field 1H NMR spectroscopy, HPLC-TOF/MS and chemometrics. Journal of Pharmaceutical and Biomedical Analysis. 2004;35(3):599-608. doi: 10.1016/j.jpba.2004.02.013

[31] Tolstikov VV, Fiehn O. Analysis of highly polar compounds of plant origin: Combination of hydrophilic interaction chromatography and electrospray ion trap mass spectrometry. Analytical Biochemistry. 2002;301(2):298-307. doi: 10.1006/abio.2001.5513

[32] Castrillo JI, Hayes A, Mohammed S, Gaskell SJ, Oliver SG. An optimized protocol for metabolome analysis in yeast using direct infusion electrospray mass spectrometry. Phytochemistry. 2003;62(6):929-937. doi: 10.1016/s0031-9422(02)00713-6

[33] Utzinger U, Heintzelman DL, Mahadevan-Jansen A, Malpica A, Follen M, Richards-Kortum R. Near-infrared Raman spectroscopy for in vivo detection of cervical precancers. Applied Spectroscopy. 2001;55(8):955-959. doi: 10.1366/0003702011953018

[34] Zhang A, Sun H, Wang P, Han Y, Wang X. Recent and potential developments of biofluid analyses in metabolomics. J Proteomics. 2012;75:1079-1088. doi: 10.1016/j.jprot.2011.10.027

[35] Wang X, Zhang A, Han Y, et al. Urine metabolomics analysis for biomarker discovery and detection of jaundice syndrome in patients with liver disease. Mol Cell Proteomics. 2012;11:370-380. doi: 10.1074/mcp.M111.016006

[36] Gevi F, Meloni A, Mereu R, Lelli V, Chiodo A, Ragusa A, Timperio AM. Urine Metabolome during Parturition. Metabolites. 2020;16;10(7):290. doi: 10.3390/metabo10070290

[37] Coleine C, Gevi F, Fanelli G, Onofri S, Timperio AM, Selbmann L. Specific adaptations are selected in opposite sun exposed Antarctic cryptoendolithic communities as revealed by untargeted metabolomics. PLoS One. 2020;15(5):e0233805. doi: 10.1371/journal.pone.0233805

Chapter 3

Pharmacometabolomics: A New Horizon in Personalized Medicine

Abdul-Hamid Emwas, Kacper Szczepski, Ryan T. McKay, Hiba Asfour, Chung-ke Chang, Joanna Lachowicz and Mariusz Jaremko

Abstract

Pharmacology is the predominant first-line treatment for most pathologies. However, various factors, such as genetics, gender, diet, and health status, significantly influence the efficacy of drugs in different patients, sometimes with fatal consequences. Personalized diagnosis substantially improves treatment efficacy but requires a more comprehensive process for health assessment. Pharmacometabolomics combines metabolomic, genomic, transcriptomic and proteomic approaches and therefore offers data that other analytical methods cannot provide. In this way, pharmacometabolomics more accurately guides medical professionals in predicting an individual's response to selected drugs. In this chapter, we discuss the potentials and the advantages of metabolomics approaches for designing innovative and personalized drug treatments.

Keywords: Personalized Medicine, Pharmacometabolomics, Metabolomics, NMR, metabolites

1. Introduction

Conventional drugs are developed as standard treatments for all patients diagnosed with particular diseases regardless of any differences between those patients. Consequently, this universal approach comes with a high degree of uncertainty regarding the treatment outcome. It is well-established that individuals can be differentially affected by the same disease due to factors such as general health status, genetics, gender, diet habits, smoking, alcoholic intake, *etc.* [1, 2]. The global COVID-19 pandemic has demonstrated clearly that a single disease can have different outcomes in different people, and the choice of therapeutic strategies needs to be calibrated to an individual rather than using a standard protocol for heterogenous populations. Indeed, the increasing incidence of treatment failure, especially with life threatening diseases such as cancer relapse, evidences a need for personalized drug regimens.

Each pathological state in humans affects multiple organs/systems and leads to the perturbation of metabolites and protein concentration levels. Thus, analysis of biomarkers (such as unique metabolites or proteins) is an effective way to monitor human health [3, 4]. Biomarkers can be used for disease prediction, diagnosis, and to screen the efficacy of treatment intervention. For example, the glucose level in blood is a biomarker of diabetes and can be used to monitor drug efficacy [5–7]. **Table 1** summarizes the most prominent examples of protein biomarkers discovered recently.

Among all '-omics' approaches, metabolomics is the most effective of qualifying and quantifying the perturbation of metabolite concentrations under external and internal factors. Thus, joining metabolomics with other '-omic' sciences (e.g. genomics) is essential for a comprehensive understanding of disease onset and pathogenesis, and provides a better diagnosis and treatment.

The total number of endogenous metabolites (although it is not completely determined yet) in human bio-fluid and tissues are lower than the total number of expressed proteins, giving metabolomics an extra advantage in monitoring disease pathology. Moreover, the perturbation of metabolite levels in human bio-fluids is usually greater than that of protein concentrations, providing an easy and clearer bio-marker role [18–20]. For instance, cancer leads to changes in affected cells, which cause an up-regulation in metabolite concentration levels during carcinogenesis [21]. For example, increased lactate levels have long been associated with different types of cancer [22]. Recently, the development of computational methods, such as bioinformatics and human metabolome databases establishing large scale bio-banks and computer programs, have facilitated the employment of metabolomics in stratified medicine. Pharmacometabolomics is a new subset of the metabolomics field aiming to predict the response of an individual to a drug or to develop optimized treatment strategies based on previous knowledge of subject

Protein biomarkers	Useful for:	Ref
Apolipoprotein H, ApoCI, Complement C3a, Transthyretin, ApoAI	Prediction of recurrence-free survival in women with estrogen receptor-negative tumors	[8]
S100 calcium-binding protein B, Neuron-specific enolase, Glial fibrillary acidic protein, Ubiquitin Carboxy-terminal hydrolase-L1, Tau, Neurofilament-light	Prediction of outcome and severity in traumatic brain injury patients	[9]
S100A9, ThioredoxiN, Cadherin-related family member 2	Diagnosis (presence) of cholangiocarcinoma	[10]
TFF1, ADAM (male only), BARD (female only)	Early diagnosis of gastric cancer	[11]
Acidic nuclear phosphoprotein 32 family member B, Thrombospondin-4, Cardiac muscle troponin T, Glucocorticoid-induced TNFR-related protein, NAD-dependent deacetylase sirtuin-2	Creating new utrophin modulation strategies that could help patients with Duchenne muscular dystrophy	[12]
C-reactive protein, S100A8, S100A9, S100A12	Prognosis of the severity of rheumatoid arthritis.	[13]
S100A4, S100A8, S100A9, Carbonic anhydrase I, Annexin V	Diagnosis of urinary bladder cancer and prognosis of patient outcome.	[14]
Gelsolin, Fibronectin, Angiotensinogen, Haptoglobin	Detection of lymph node metastasis of oral squamous cell carcinoma.	[15]
Neurotrophic factor, Angiotensinogen, Insulin-like growth factor binding protein 2, Osteopontin, Cathepsin D, Serum amyloid P component, Complement C4, Prealbumin (transthyretin)	Diagnosis of Alzheimer's disease in Han Chinese.	[16]
Alpha-2-macroglobulin, Chromogranin-A, Glutathione pertidase 3	Obtaining qualitative and quantitative assessments of radiation exposure.	[17]

Table 1.
Examples of biomarkers and their use in medicine.

DOI: http://dx.doi.org/10.5772/intechopen.98911

Year	Number of approved novel drug by FDA
1993	25
1994	21
1995	29
1996	53
1997	39
1998	30
1999	35
2000	27
2001	24
2002	17
2003	21
2004	36
2005	20
2006	22
2007	18
2008	24
2009	26
2010	21
2011	30
2012	39
2013	27
2014	41
2015	45
2016	22
2017	46
2018	59
2019	48
2020	53

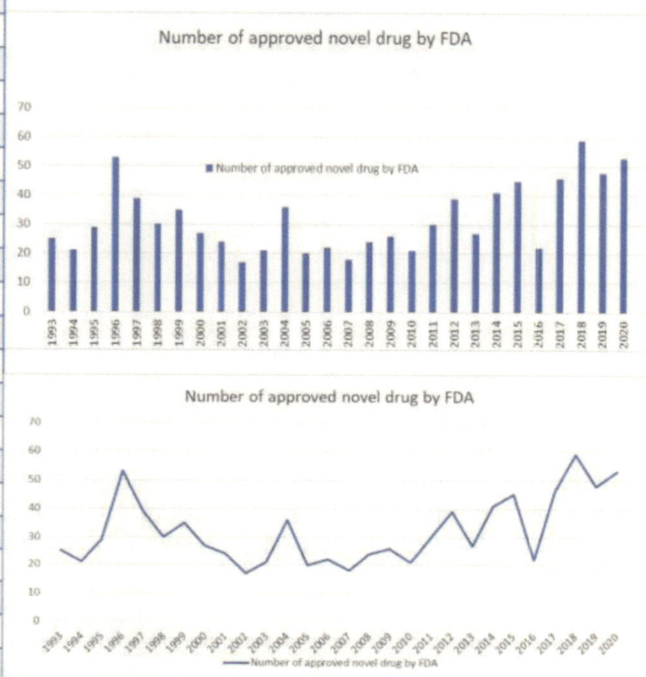

Figure 1.
Number of novel drugs approved annually by the FDA between 1993 and 2020 with graphical representation.

metabolomics information (individual's metabolic profile). One should keep in mind that aerosolized treatment would never lead to the discovery of a novel drug for each individual subject. Indeed, the number of new drugs is almost constant in the last decades (**Figure 1**).

In this chapter, we briefly introduce metabolomics along with common metabolomics analytical platforms regarding the development of a personalized medicine approach and factors that will empower advances in personalized medicine.

2. FDA approved drugs since 1975

Over the past few decades, pharmaceutical product intervention has improved significantly resulting in more saved lives and enhanced public health. The annual number of newly approved drugs applicable for human use has varied greatly over the years (**Figure 1**). The Food and Drug Administration (FDA) is an agency within

Years	Average FDA novel drug approval per year
1970s	10
1980s	10
1990s	20-25
2000-2010	25.6
2010-2020	39.2

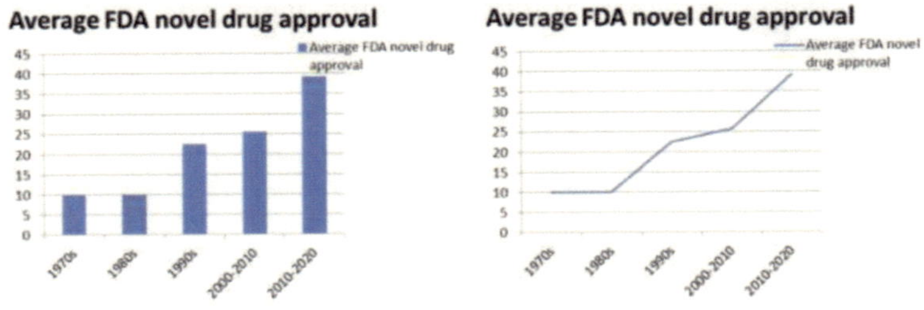

Figure 2.
Average numbers of novel drug approvals by the FDA over the last five decades with graphical representation.

the United States. One of its primary responsibilities is the approval of human pharmaceutical products based on safety and efficacy. Regulating and managing the human pharmaceutical industry and the approval of new drugs is the responsibility of the Center for Drug Evaluation and Research (CDER) [23].

The FDA catalog contains most of the approved drug products since 1939. However, since 1998 a complete human drug database is available, known as the Orange Book, which includes patient information, drug labels, and drug reviews. The Orange Book is considered a comprehensive detailed list of all pharmaceutical products approved in the U.S. by the FDA. However, studying the number of pharmaceutical products approved annually is not straightforward. First, the number of approved human drugs was not accurate before 1981, as the Orange Book did not report pharmaceutical drug approval data until after 1981, including new molecular entities (NMWE), the pharmaceutically active ingredient, drug dosage form, combination, formulation, and indication [24]. In addition, the Orange Book excludes any withdrawn drug or 'no-longer marketed' pharmaceutical products due to either drug efficacy concerns or safety concerns. Below, the reader can find **Figure 1** summarizing the number of FDA-approved novel human drugs per year from 1993 to 2020 [24, 25].

As is apparent from **Figure 1**, the year 2020 represents the second-highest number of FDA-approved novel human drugs over the past twenty years (53 drugs), while 2018 was the year when the highest number of drugs were approved by the FDA (57 drugs). In 2017, only 46 drugs were approved [26].

The average rate of new drug approvals by the FDA has increased over the years (**Figure 2**). Before 1950, the average annual drug approval was less than four, while the average annual drug approval in the 1960s and 1970s was 10. However, in the 1980s the average approval rate increased to more than twenty per year. It has continued to increase to reach more than twenty-five approvals per year from 2000 to 2010. Over the last several years there have been further increases, reaching an average of more than 39 approved compounds per year from 2010 till 2020 [26, 27]. The average novel drug approval by the FDA over the decades is listed in **Figure 2**.

3. Metabolomics

Metabolomics is defined as "the measurement of metabolite concentrations and fluxes and secretion in cells and tissues in which there is a direct connection between the genetic activity, protein activity, and the metabolic activity itself" [28]. It is a relatively new field and is employed in a wide range of applications that monitor biological systems [3, 29, 30]. Integrating metabolomics with other '-omics', including proteomics, transcriptomics, and genomics, provides an exhaustive description of the biological system under study. Metabolomics provides a snapshot of the metabolite dynamics, and is a powerful tool when investigating numerous perturbations in biological systems, including pathophysiological events, environmental stimuli, and genetic modifications [31–34]. Moreover, metabolomics investigates every perturbation in metabolite compositions and/or concentrations, and it has already been applied in different fields such as biomedicine, environmental science, nutrition and diet studies, microbiology, and drug toxicology, as well as marine and plant sciences [35–39].

Metabolomics is usually classified into two main categories: (1) untargeted, and (2) targeted. Untargeted metabolomics is focused on the entire pool of "detectable" metabolites in a biological sample and makes no assumptions about metabolite(s) or class of metabolites, nor their concentrations. Untargeted metabolomics relies on fingerprinting approaches, where a group or different classes of samples (e.g., healthy control vs. pathological samples) are compared, and where absolute metabolite quantifications are not necessary. In contrast, targeted metabolomics focuses both on the identification and quantification of a specific number of metabolites. Targeted metabolomics approaches are relevant for drug development, where the drug mechanism (including drug absorption and drug distribution) needs to be precisely monitored.

The choice of proper analytical technique(s) in metabolomics is the crucial step, and particularly targeted metabolomics requires accurate metabolite quantification. Metabolomics applies different analytical techniques, including mass spectrometry (MS), nuclear magnetic resonance (NMR) spectroscopy, Fourier transformed infrared (FT-IR) spectroscopy, and high-performance liquid chromatography (HPLC). Among them, NMR spectroscopy and MS spectrometry are the most common and powerful analytical tools [40, 41].

3.1 Analytical techniques in metabolomics

Similar to other '-omics' disciplines, metabolomics uses different analytical platforms, separately or in combination (two or more techniques) [32, 42]. Although several analytical platforms are employed in metabolomics studies, including FT-IR spectroscopy [43–45], HPLC [46, 47], NMR spectroscopy [48–53], and MS [54–57] combined with gas or liquid chromatography [58–62], MS and NMR are the most common approaches [3, 50, 63–65]. There is no single optimum analytical technique that can elucidate all metabolites equally. Each analytical method has its advantages and limitations. For example, NMR is a non-destructive and highly reproducible technique where metabolic pathways or metabolic flux can be studied by using isotopic nuclei (such as ^{13}C and ^{15}N NMR), thus monitoring the flow of compounds through metabolic pathways [66–69].

Nevertheless, it has two main drawbacks that must be kept in mind: inherently low sensitivity and potential signal overlap. Different technical approaches have been developed to overcome these two drawbacks, contributing to the development of new and more efficient NMR techniques. For example, dynamic nuclear polarization (DNP) can be used to increase the NMR signal enhancement [70, 71], and the use of cryoprobes and the introduction of ultra-high magnetic fields (e.g., 1GHz) helps to overcome the low sensitivity limitation [72, 73]. The peak overlap problem can be minimized by the use of the highest magnetic fields and multi-dimensional NMR methods such as HSQC, TOCSY, COSY, and HMBC techniques [66, 74–78].

As stated, no singular analytical technique can perform a complete quantification and identification of all the metabolites in one analysis. Therefore, in addition to one and two-dimensional NMR experiments, different complementary techniques are required, such as liquid chromatography-mass spectrometry (LC–MS) and gas chromatography–mass spectrometry (GC–MS), which help to maximize the number of identified and quantified metabolites [32, 65, 79, 80].

For instance, the human urine metabolome was analyzed by Wishart *et al.* with several different analytical tools (ICP-MS, NMR, GC–MS, DFI/LC–MS/MS, HPLC) to facilitate the detection of the highest possible number of human urine metabolites. Among all metabolites, 209 were identified by NMR, 179 by GC–MS, 127 by DFI/LC–MS/MS, 40 by ICP-MS, and 10 by HPLC [81].

Based on the ability to separate and detect a wide range of metabolites, LC–MS is one of the most widely used tools for carrying out metabolite profiling studies [82–86]. LC–MS combines HPLC and mass spectrometry, and provides a powerful analytical tool for the separation, identification, and quantification of metabolites in a studied sample [65, 87–90]. HPLC separates molecules based on different physical and chemical properties such as charge, polarity, molecular size, and affinity towards column matrices [91–94]. Thus, different successful chromatography methods have been developed, such as reversed-phase (RP) gradient chromatography [85, 86, 95, 96]. To obtain the best separation, and presumably the highest number of detected metabolites, each sample can be analyzed twice using RP and normal phase chromatography. Moreover, the column switching approach of 2-dimensional analysis in an "orthogonal" combination of hydrophilic interaction liquid chromatography (HILIC) and RP-L, in conjunction with utilizing different electro spray ionization (ESI) modes can also be used [85, 86, 97–99]. In addition to using different separation methods and/or ionization methods, LC–MS is inherently far more sensitive than NMR and enables researchers to detect secondary metabolites at lower concentrations [100, 101]. The drawback occurs with the consistency of the separation performance. For example, columns can degrade non-linearly over time, requiring constant monitoring, determination of effect(s), and compensation in the final analysis. Solvent purity, pump performance, and injector consistency can all come into play. The inclusion of quality control samples at the beginning, end, and randomly inserted into the experimental samples should allow the compensation and quality control of any introduced confounder(s), but adds material costs, extends batch run times, and introduces complexity to the analyses.

3.2 Development of '-omics' in personalized medicine approach

Over the last decades, various fields of bioresearch (genetics, genomics, proteomics, and metabolomics) have quickly evolved and revealed mechanisms of diseases, and most importantly delivered new therapeutic outcomes. Although the current tenet regarding the uniformity of the drug response seems to be widely accepted, it does not take into account the individual differences. Individuals may not respond in the same way to the pharmacological treatments or present minor and serious side effects. For example, antidepressants [102], statins [103, 104], or antipsychotic drugs [105] have been shown to have reduced effects on some individuals, even to the extent that only a quarter of patients can achieve a functional remission of the disease [105]. Pharmaceutical treatments are ineffective for 30 to 60% of patients [106]. Moreover, a significant number of patients may develop adverse drug reactions (ADR) related to their treatment, with the incidence of fatal ADR being 0.32% [107]. In order to minimize the negative effects of pharmaceutical treatments, and at the same time optimize the drug therapy in terms of its efficiency, a more personalized approach has been proposed, which assesses various factors prior to the treatment through the application of the different '-omics' [108].

This approach is not entirely new, as some characteristics (age, weight, co-morbidity, family history, and biochemical parameters) are already commonly considered. However, technological progress allows us to analyze individuals in more detail – from different genes, and single-nucleotide polymorphism (SNPs) genomics, to small, biologically active molecules (proteomics, metabolomics) and even the metabolic pathways of individuals (metabolomics, fluxomics) [109, 110]. In addition, personalized medicine not only takes into account the physiological status of a person's body - it also considers the unique, psychosocial situation of the individual, which may have a direct effect when a given health condition manifests in that individual and how he/she will respond to treatment [111]. Although these aspects are taken into consideration for a more complete picture of a person's medical status, separate approaches could also be used to focus on precise problems. For example, a fairly new field called pharmacogenomics tries to assess and validate the impact of human genetic variation on drug responses [112, 113]. Currently, we know that inherited variations in approximately 20 genes can affect around 80 medications and the way the body responds to them [114]. Another young field that has become a prominent branch of metabolomics is pharmacometabolomics, which is the subject of this review.

Personalized medicine has already shown its value in therapies to combat diabetes and cancer [115–119]. For example, the management of blood glucose in diabetes requires proper medication, for which the dosage and efficiency is suited to the individual patient. The efficacy of one of the drugs used in type 2 diabetes, metformin, has been associated with polymorphisms in several genes, specifically solute carrier family (SLC) 22 (an organic cation transporter) member 1 (*SLC22A1*), *SLC22A2*, *SLC47A1*, organic cation transporter 1 and 2 (*OCT1* and *OCT2*), and the gene encoding for multidrug and toxin extrusion 1 protein [MATE1] [115, 120]. Sulfonylureas which are another class of drugs used to treat type 2 diabetes, have been shown to have a variable response effect depending on the genomic profile of the patient, *e.g.*, the variant 'TT' at rs12255372 in the *TCF7L2* gene results in a weaker response compared to the 'GG' version [116, 121]. Those genetic factors are usually not considered when therapy is administered, despite the fact that the information they provide can have direct and substantial effects on therapy optimization and the success of treatment.

Similar benefits from personalized medicine have been observed in the treatment of various types of cancer. One of the best examples that highlights recent progress is breast cancer. Based on the biomarkers present in tumors, such as the estrogen receptor, progesterone receptor, antigen Ki-67, human epidermal receptor 2 [122], and mutations in genes such as Breast cancer gene 1 and 2 (*BRCA1*, *BRCA2*), which are related to carcinogenesis [123], breast cancer can be divided into different subgroups [122]. Each of the cancer types has its own characteristics and requires a specific, more personalized approach to maximize treatment efficacy while minimizing the adverse effects [122, 124]. The decision regarding which therapy to choose becomes even more complicated when we also consider the genetic profile of an individual (the susceptibility to the treatment) [118, 122, 125]. For example, different variants of CYP2D6 (cytochrome P450, family 2, subfamily D, polypeptide 6), which interacts with tamoxifen (a standard drug used in steroid receptor positive breast cancer) have been shown to have direct impacts on the treatment. The impaired version of the protein could also be associated with the recurrence of breast cancer [118, 122]. On the other hand, a personalized approach could also be used in a preventive way. As an example, genetic testing with a focus on the identification of potential, carcinogenic mutations in the *BRCA1* and *BRCA2* genes could be used to create a proactive strategy (MRI, chemoprevention, bilateral mastectomy), thus significantly decreasing the chances of developing a more severe disease [126].

3.3 Metabolomics databases

The demand for functional and inclusive metabolomics databases is driven by the need for fast data analysis including metabolite identification, quantification, and subsequent interpretation of complex metabolite data, and possibly from multiple instrument sources. As a result of collective efforts in this area, several different databases have been established, including the Human Metabolome Database (HMDB) (https://hmdb.ca/) [127, 128], Platform for RIKEN Metabolomics (PRIMe) [129], Biological Magnetic Resonance Data Bank (BMRB) [130], and the Madison Metabolomics Consortium Database (MMCD) [131]. The existing information on the human urine metabolome was published recently with detailed information on each reported metabolite, including concentration perturbation at normal and disease-related levels (http://www.urinemetabolome.ca). The human urine metabolome along with the human serum metabolome represent a significant development and resource for researchers, which may be critical when employing metabolomics approaches in clinical applications including stratified medicine. Furthermore, the human metabolome database serves as a cross-referencing and benchmarking tool for general metabolomics studies, including metabolite identification, quantification, and newly discovered disease biomarkers. The Madison-Qingdao Metabolomics Consortium Database (http://mmcd.nmrfam.wisc.edu/) contains information on more than 20,000 compounds, including NMR and MS data that are valuable in the identification and quantification of metabolites present in biological samples [131]. Among different freely available metabolomic databases, the HMDB (www.hmdb.ca) [128, 132] (University of Alberta, Canada, David Wishart group) is becoming the de-facto standard reference for the metabolomics community. The HMDB contains information on 74,462 metabolite entries gathered and summarized from literature-derived data and also contains an extensive collection of experimental metabolite concentration information compiled from hundreds of MS and NMR metabolomics analyses performed on urine, blood, and cerebrospinal fluid samples. The data entries encompass a wide range of information, including structural, chemical, clinical, and biological information for many of the reported metabolites.

In 2012, the MetaboLights database (http://www.ebi.ac.uk/metabolights) [133] was established for the online storage of metabolomic experiments, associated raw data, and metadata, to interrogate databases of collected experimental information in publications. This database was first developed and maintained by the European Bioinformatics Institute [134], and later it has been endorsed and developed by the COSMOS consortium [135]. The continuous development of metabolomics databases alongside the uninterrupted advancements in software and supercomputer capabilities may lead to better clinical practices, including diagnosis, disease prognosis, and, ultimately, effective personalized treatments.

3.4 Biobanks and their impact on personalized medicine studies

Over the past decade, several high-capacity biobanks have been established to serve as baseline research and clinical studies tools in use by scientific institutions, clinics, private companies, and regulators at both national and international levels, encompassing a high number (*i.e.*, millions) of samples necessary for medical research. Furthermore, the standardization of sample collection and storage conditions will help reduce sample collection bias and overcome the limitations afforded by variations between different studies, protocols, and practices. Biobanks usually also collect relevant data, such as whole-genome, genotype, geographic location, dietary preference(s), proteomic, and medical image information [136–138]. In addition to national registries, the incorporation of existing electronic health records (EHRs) is becoming more common, making large biobank datasets more applicable for a greater number of users [139, 140]. The availability of this additional information, combined with the collection of multiple samples over longer periods from the same individual, facilitates improved interpretation of experimental data and provides controls for possible confounders. Establishing large scale national and international biobanks therefore is an essential step and a valuable resource for clinical practitioners and in the development of public health policies, in addition to being crucial for the development of personalized treatments. These megabanks have the capacity to store samples from the same person over the course of many years, which in the future may be collected from childhood and followed up with the periodic collection of new samples throughout life [141].

As biobanks represent a major resource in large-scale global studies, we believe that the impact of metabolomics approaches will become ever more important in future medical research and public health efforts, including personalized health care and stratified medicine.

3.5 Pharmacometabolomics

As mentioned, pharmacometabolomics is a fairly new addition to the '-omics" family of studies. One of the pioneering works that helped create this novel field of science was carried out in 2006 by Clayton *et al.* on paracetamol [142]. Their main goal was to check if the metabolite profile of an animal, prior to the administration of a drug, would allow for the prediction of the metabolism of a drug as well as its toxic effects on an animal. For this purpose, the team collected urine samples from 65 rats, both before and after the administration of paracetamol. Later, samples were analyzed by ^{1}H 1D NOESY NMR spectroscopy. After analyzing the spectra, researchers identified four paracetamol-related metabolites, specifically paracetamol sulphate, paracetamol glucuronide, mercapturic acid derived from paracetamol, and paracetamol. Compared to the histopathological results obtained from the same rats, a substantial model for predicting post-dose histology of the liver could not be established, but they discovered a relationship between the pre-dose metabolic profile of urine and the histological outcome. The main factors predicting that relationship were identified as taurine, trimethylamine-N-oxide (TMAO), and betaine, where higher pre-dose levels of taurine indicated less damage to the liver while higher levels of TMAO and betaine were associated with greater damage [142]. This pioneering work paved the way for the establishment of organizations focused on pharmacometabolomics, such as the Pharmacometabolomics Research Network (PMRN), where the main aim is to "integrate the rapidly evolving science of metabolomics with molecular pharmacology and pharmacogenomics" [143]. So far, PMRN has produced numerous publications, many of them pertaining to lifestyle disorders and diseases. One example concerns the lipidomic response to treatment with simvastatin [144]. The authors of this paper identified metabolites that could predict the outcome of treatment with simvastatin – phosphatidylcholine, including 18 carbon fatty acids with two double bonds at the n6 position, cholesterol esters with 18 carbon fatty acids with one double bond at the n7 position, and 18 carbon-free fatty acids with three double bonds at the n3 position [144]. Additionally, the authors discovered a group of metabolites that may help to predict the changes of C-reactive protein (CRP) after the treatment - five of them were plasmalogens (a specific group of glycerophospholipids containing a vinyl ether moiety at the sn-1-position of the glycerol backbone) [145], and the others were phosphatidylcholines and cholesterol esters [144]. Another interesting study worth mentioning is related to changes in lipids levels in schizophrenia and upon treatment with antipsychotics [146]. The authors measured the changes in the lipid profiles of patients before and after treatment with olanzapine, risperidone, and aripiprazole. They discovered that treatment

Type of Biomarker	Definition	Example
Diagnostic Biomarker	Biomarker used to detect or confirm the presence of disease or to identify individuals with a subtype of the disease.	Sweat chloride can be used to confirm cystic fibrosis [148].
Monitoring Biomarker	Biomarker measured constantly to assess the status of the disease or for evidence of exposure to (or effect of) a medical product or an environmental agent.	HIV-RNA can be used as a monitoring biomarker to measure and guide treatment with antiretroviral therapy (ART) [149].
Pharmacodynamic/Response Biomarker	Biomarker used to show a biological response of an individual exposed to a medical product or an environmental agent.	Serum LDL cholesterol can be used for evaluating response to a lipid-lowering agent in patients with hypercholesterolemia [150].
Predictive Biomarker	Biomarker used to identify individuals who will experience positive or negative outcome from exposure to a medical product or an environmental agent.	Mutations in BRCA 1/2 genes can be used to identify women with platinum-sensitive ovarian cancer that will most likely respond to PARP inhibitors [151].
Prognostic Biomarker	Biomarker used to identify the likelihood of a clinical event such as disease recurrence or progression.	Mutations in BRCA 1/2 genes can be used to evaluate the likelihood of a future second breast cancer in patients currently diagnosed with one [152].
Safety Biomarker	Biomarker used for indicating the likelihood or presence of a toxic effect, measured before or after the exposure to a medical product or an environmental agent.	Hepatic aminotransferases and bilirubin can be used to evaluate potential hepatotoxicity [153]
Susceptibility/Risk Biomarker	Biomarker used for the estimation of a chance of disease or other medical condition in an individual who currently does not have clinically apparent disease or condition.	Mutation in BRCA 1/2 genes can be used to identify individuals with a predisposition to develop breast cancer [154].

Table 2.
Types of biomarkers with examples of practical applications. Based on the BEST (Biomarkers, EndpointS, and other Tools) resource by the FDA-NIH Biomarker Working Group [147].

with olanzapine and risperidone increased the levels of 50 lipids, raised the concentration of triacylglycerols and generally decreased free fatty acids. Moreover, the concentration of phosphatidylethanolamine that is suppressed in patients with schizophrenia was raised by all three drugs [146].

Presently, most of the pharmacometabolomics studies focus on identifying specific biomarkers related to administered medications. Those biomarkers can provide information ranging from predicting patient treatment response, to monitoring the changes during the treatment, or evaluating the end effects of treatment (i.e., if the patient responded positively or negatively to the therapy) (see **Table 2** and **Figure 3**). Examples of pharmacometabolomic studies are shown in **Table 3**.

The successful isolation of a metabolite that may become a biomarker depends on the type of sample and the approach. In addition to easily and commonly accessed samples like urine and blood serum, pharmacometabolomics studies can also utilize feces, saliva, human breast milk, and even breath [175–177]. Samples are usually collected before, during, and after the treatment, and can be further divided by type of response from an individual (*e.g.*, mostly positive, mostly negative, or intermediate) [175, 178]. After obtaining data from a set of samples using various techniques adapted to the particular type [36, 175, 178], a database for each individual is created, with metabolites detected and identified before and after the treatment [178]. Lastly, a statistical analysis is applied to obtain information ranging from differences that can distinguish good and poor responders prior to the treatment, to changes in metabolites due to drug application that can be correlated with response phenotypes and assumptions of pathways connected to variants of response [178].

For example, Wikoff and colleagues [179] investigated atenolol-induced changes in Caucasians and African Americans. Atenolol is a beta-adrenergic receptor blocker used in a first line antihypertensive treatment. However, various patients responded quite differently. The main objective of this study was to obtain metabolic signatures of atenolol treatment that provided insight into racial differences in response to beta blockers. They found that atenolol has a strong impact on fatty acids in blood serum, but the results were different for different groups (*e.g.*, effects of treatment were highly significant in Caucasians but minimal in African Americans). Furthermore, the authors examined associations between oleic acid and SNPs on the 16 genes encoding lipases. They discovered that a SNP in the *LIPC* (rs9652472) and *PLA2G4C* (rs7250148) genes were associated with the change in oleic acid concentration in Caucasians and African Americans, respectively [179]. Another example of utilizing a combined approach

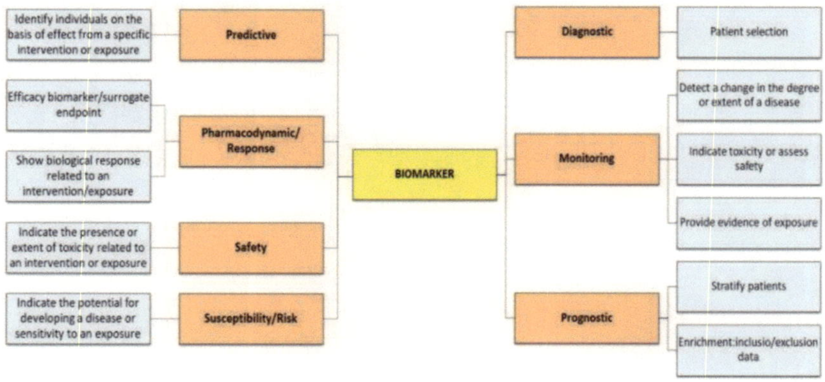

Figure 3.
A brief description of biomarkers of specific use in the drug development process. Based on "Context of use (COU) for a biomarker" by U.S. Food and Drug Administration [155].

Chemical Compound	Goal of the study	Main changes in metabolites post exposure	Conclusions	Ref
Aspirin	To investigate the mechanism of aspirin resistance.	↑ Inosine, adenosine, guanosine ↓ Hypoxanthine, xanthine	• Higher levels of adenosine and inosine were observed in the group categorized as "poor responders". • A pharmacogenomics approach pinpointed an SNP in the adenosine kinase (ADK) intronic variant - rs16931294, where the G allele of this variant was associated with poor response to the treatment.	[156]
	To define pathways implicated in variation of response to treatment with a focus on metabolites containing an amine functional group.	↑ O-Phosphoethanolamine, serotonin ↓ Glycylglycine, L-aspartic acid, L-glutamic acid, L-leucine, L-phenylalanine, L-serine, ethanolamine, glycine, ornithine, taurine, L-asparagine, L-valine, beta-alanine, L-lysine, L-histidine, L-tyrosine, L-glutamine	• The changes in metabolite profiles of healthy individuals treated with low dosage of aspirin cannot be directly attributed to COX-1 inhibition. • Increased levels of serotonin correlated with higher post-aspirin platelet reactivity.	[157]
	To investigate: • The effects of low-dose aspirin therapy on the oxylipid metabolic pathways, • the sex differences in aspirin-induced oxylipid changes, and • potential association of oxylipid on aspirin-induced inhibition of platelet aggregation.	↑ 13,14-dihydroPGF2 ↓ TXB2, 12-HHTrE, 11-HETE, 5-HETE, 12-HETE, 8-HETE, 15-HETE, 9-HODE, 13-HODE, 5-HETrE, 5-HEPE, 12-HEPE, 15-HEPE, 9-HOTrE, EpOMEs, DiHOMEs, DiHETrEs, 20-HETE.	• Aspirin does not show any sex-specific effects on oxylipid levels. • Aspirin decreases almost all of the oxylipids measured in the samples. • Several LA-derived oxylipid (3-HODE, 9-HODE, 12,13-DiHOME, and 12,13-EpOME) metabolites might contribute to the variability of non-COX1-mediated response to aspirin.	[158]
	To assess the metabolic pathways affected by aspirin administration that are potentially involved in cardiovascular and antitumoral protection.	↑ 3-methylglutarylcarnitine ↓ L-histidine, hydantoin-5-propionate, 4-imidazolone-5-propanoate, N-formimino-L-glutamate, xanthosine, L-glutamine, 5-aminoimidazole-4-carboxamide-1-β-D ribofuranoside, butyryl-L-carnitine, tiglylcarnitine, isovalerylcarnitine, heptanoylcarnitine,	• Aspirin decreases the levels of glutamine and metabolites involved in histidine and purine metabolism. • The ability of aspirin to increase the β-oxidation of fatty acids and decrease glutamine levels implicates reduced synthesis of acetyl-Co-A that could help explain aspirins potential anticancer effects.	[159]

DOI: http://dx.doi.org/10.5772/intechopen.98911

Chemical Compound	Goal of the study	Main changes in metabolites post exposure	Conclusions	Ref
Aspirin eugenol ester (AEE)	To evaluate the protective effect of AEE on paraquat-induced acute liver injury (ALI) in rats.	↑ L-histidine, D-asparagine, L-phenylalanine, pipecolic acid, acetylglycine, N-(2-methylpropyl)acetamide, inosine, xanthosine, melatonin radical, ophthalmic acid, glutamylarginine, S-(PGJ2)-glutathione, L-octanoylcarnitine, lysoPC(P^{-16}:0), argininic acid, N-undecanoylglycine, chenodeoxyglycocholic acid, ↓ Glycerophosphocholine, hypoxanthine, nonyl isovalerate, glutamylleucine, pipecolic acid, deoxycholic acid glycine conjugate, dephospho-CoA, taurochenodesoxycholic acid, lysoPC(14:1), PA(22:2), cholic acid, 5,9,11-trihydroxyprosta-6E,14Z-dien-1-oate, lysoPE(18:2), lysoPE(20:4), lysoPE(16:0)	• AEE shows protective effects against PQ-induced ALI. • The mechanisms in which aspirin eugenol ester protects against the effects on PQ-induced ALI are correlated with antioxidants that regulate amino acid, phospholipid, and energy metabolism metabolic pathway disorders and attenuate liver mitochondria apoptosis.	[160]
	To identify the different proteins and small molecules in plasma to explore the mechanism of action of AEE against thrombosis.	↑ Oleamide, palmitic amide, linoleic acid, L-acetylcarnitine, creatine, proline betaine, arachidonic acid ↓ L-carnitine, L-methionine, L-proline, L-pipecolic acid, allantoin, palmitic acid, citric acid, L-tryptophan*	• Metabolomics results suggested that the therapeutic mechanism of action of AEE (as well as for aspirin and eugenol) could be involved with energy metabolism, amino acid metabolism, and fatty acid metabolism. • A total number of 38 (AEE), 41 (aspirin) and 54 (eugenol) proteins were differentially regulated in rats treated with those compounds.	[161]
Busulfan	To investigate biomarkers for predicting busulfan optimal dosage.	↑ Deferoxamine-derived metabolites ↓ Carnitine C9:1, carnitine C12:1-OH, phenylacetylglutamine**	• Busulfan metabolism is decreased in patients with high ferritin levels and reduced liver function.	[162]
Gemcitabine	To investigate potential predictive biomarkers for the efficacy of gemcitabine-based chemotherapy while obtaining the most optimal therapeutic results in patients with pancreatic cancer.	A total number of 38 and 26 different metabolites were identified between the gemcitabine resistant and gemcitabine sensitive pancreatic carcinomas from whom four of them: 3-hydroxyadipic acid, D-galactose, lysophosphatidylcholine (LysoPC) (P-16:0) and tetradecenoyl-L-carnitine, were significantly different between the carcinoma types.	• 3-hydroxyadipic acid, D-galactose, lysophosphatidylcholine (LysoPC) (P-16:0) and tetradecenoyl-L-carnitine could be used as biomarkers for evaluating the efficacy of chemotherapy in pancreatic carcinoma.	[163]

Chemical Compound	Goal of the study	Main changes in metabolites post exposure	Conclusions	Ref
Isoniazid (INH), Rifampicin (RIF), Pyrazinamide (PZA), Ethambutol (EMB) - DOTS treatment program	To identify metabolites that describe the changes related to tuberculosis therapy	↑ Dodecyl acrylate, pyrazinamide, 1,6-hexylene glycol, ribitol, 1-decene, 2,4-dimethylbenzaldehyde, 2,6-dimethylnonane, 3,4-dihydroxybutyric acid, 5-hydroxyindoleacetic acid, alfa-isosaccharinic 1,4-lactone, beta-Isosaccharinic 1,4-lactone, decane, fumaric acid, hippuric acid, N-formylglycine, sebacic acid, threonic acid, undecane, urea, 3-ethyl-4-methyl-1Hpyrrole-2,5-dione, D-lyxose, phosphoric acid, ↓ Pyrazinoic acid, ethylene glycol, oleic acid, 5-oxoproline, citric acid, ethyl ester, cumene, hemimellitene, hexadecane, indane, isocumene, o-ethyltoluene, oxalic acid, p-ethyltoluene, sorbose, vannilic acid, cyclobutanamine***	• Metabolite markers that are associated with oxidative stress decline between weeks 2 and 4 of treatment – a sign of patient recovery. • During the tuberculosis therapy several enzymes (CYP2E1, CYP3A4, alcohol dehydrogenase, aminocarboxymuconate-semialdehyde decarboxylase) undergo inhibition in a time-dependent manner. • During treatment, the urea cycle is upregulated, and the production of insulin is altered.	[164]
Paclitaxel	To investigate the association between pretreatment metabolome, early treatment-induced metabolic changes, and the development of paclitaxel-induced peripheral neuropathy for breast cancer patients.	↑ Pyruvate, alanine, threonine, phenylalanine, tyrosine, asparagine, lysine, o-acetylcarnitine, proline, lactate, glutamine, leucine ↓ 3-hydroxy-butyrate, 2-hydroxybutyrate****	• Pre-treatment levels of histidine, phenyl-alanine, and threonine may predict severity of potential peripheral neuropathy.	[165]
	To investigate metabolite signatures prior to the treatment, in order to explain the variability of paclitaxel-induced pharmacokinetics.	↑ Creatinine, glucose, lysine, lactate ↓ Betaine	• Pre-treatment levels of creatinine, glucose, lysine, lactate and betaine could be associ-ated with variability of paclitaxel-induced pharmacokinetics	[166]

Chemical Compound	Goal of the study	Main changes in metabolites post exposure	Conclusions	Ref
Irinotecan	To identify metabolite changes that could have potential implications on the mechanism of action of irinote and could serve as biomarkers for efficiency of a treatment.	↑ N-α-acetyllysine, 2-aminoadipic acid, asymmetric dimethylarginine, cystathionine, propionylcarnitine,L-acetylcarnitine, malonylcarnitine, valerylcarnitine, thymine, uracil, xanthine	• The increased levels of purine and pyrimidine nucleobase metabolites could be the result of purine/pyrimidine nucleotide degradation (break of double stranded DNA in cancer cells) as a response to the treatment with irinotecan. • The increased levels of acylcarnitines and amino acid metabolites could reflect dysfunction of mitochondria and oxidative stress in the liver.	[167]
Docetaxel (DTX)	To evaluate the response of MCF7 tumor cells to high (5uM) and low (1 nM) doses of DTX.	For high dosage (5uM): ↑ Phosphoethanolamine, cytidinediphosphocholine, polyunsaturated fatty acid, ↓ Phosphatidylcholine, glycerophosphocholine, glycerophosphoethanolamine, total glutathione, glutamate, arginine, lysine, lactate, acetate, For low dosage (1 nM): ↑ Phosphoethanolamine, cytidinediphosphocholine, homocysteine, aspartate, ↓ Phosphatidyl-choline, glycerophosphocholine, hypotaurine, taurine, total glutathione, arginine, alanine, threonine, lysine, acetate,	• Both dosages result in inhibition of phosphatidylcholine biosynthesis and decreased levels of glutathione. • The mechanisms responsible for decreased glutathione levels are different. At high dosage, the extensive consumption and precursor starvation was the main reason, while for low dosage, it was the inhibition of trans-sulfuration that inhibited glutathione biosynthesis.	[168]
Metformin	To identify urinary markers of metformin responses in patients with type 2 diabetes mellitus.	↑ Myoinositol, hypoxanthine ↓ Citric acid, pseudouridine, p-hydroxyphenylacetic acid, hippuric acid*****	• Citric acid, myoinositol and hippuric acid have the potential to become biomarkers that could predict the response to metformin in patients with type 2 diabetes mellitus.	[169]

Chemical Compound	Goal of the study	Main changes in metabolites post exposure	Conclusions	Ref
Simvastatin	To investigate the metabolic changes connected with the increased risk of developing hyperglycemia as an adverse response to simvastatin.	↑ Glucose, glutamic acid, alanine, ↓ Lauric acid, myristic acid, linoleic acid, glycine, palmitoleic acid, 3-hydroxybutanoic acid, aminomalonate, oleic acid, N-methylalanine******	• Patients showing a mild resistance to insulin tend to develop full insulin resistance after simvastatin treatment. • Branched-chain amino acids, and other metabolites such as ketoleucine, hydroxylamine and ethanolamine could predict type 2 diabetes mellitus risk following simvastatin therapy.	[170]
Olanzapin	To reveal the pharmacodynamics and mechanism of action of olanzapine.	↑ Tyrosine, succinic acid semialdehyde, homovanillic acid, 3,4-dihydroxyphenylacetic acid, L-asparagine ↓ 5-hydroxytryptamine, –5- hydroxyindoleacetic acid, L-3,4-dihydroxyphenylalanine, γ-aminobutyric acid, kynurenine, kynurenine acid, tryptophan, glutamic acid, taurine, acetylcholine	• Olanzapin alters glycerophospholipid metabolism, sphingolipid metabolism and the citrate cycle.	[171]
Losartan	To predict inter-individual variations in the metabolism of losartan.	↑ Lipid CH3 (LDL/VLDL), lipid CH2 (LDL), lactate, citrate, creatine, α-glucose ↓ Lipid CH3 (HDL), creatinine, choline, glycine, phosphorylcholine	• Identification of 11 potential biomarkers from whom lactic acid, creatinine, glucose, and choline showed a good score for prediction of metabolic processes of losartan.	[172]
Midazolam, Ketoconazole, Rifampicin,	To predict biomarkers related to midazolam sum of the clearance related to the induction and inhibition of CYP3A.	↑ 6β-hydroxycortisol/cortisol, 6β-hydroxycortisone/cortisone, 16α-hydroxy-DHEA/DHEA, 16α-hydroxyandrostenedione/androstenedione, 4-hydroxyandrostenedione/androstenedione, 7β-hydroxy-DHEA/DHE,6β-hydroxyandrostenedione/androstenedione, 2-hydroxyestrone/estrone, 2-hydroxyestradiol/estradiol, 11β-hydroxyandrosterone/androsteron, 11β-hydroxyandrostenedione/androstenedione ↓16α-Hydroxytestosterone/testosterone, 11β-Hydroxytestosterone/testosterone******	• Urinary DHEA levels, 7β-hydroxy-DHEA:DHEA ratios, 6β-hydroxycortisone: cortisone ratios could be used to predict sum of the clearance for midazolam	[173]

DOI: http://dx.doi.org/10.5772/intechopen.98911

Chemical Compound	Goal of the study	Main changes in metabolites post exposure	Conclusions	Ref
DA-9701 (extract from *Pharbitis nil* seed and *Corydalis yanhusuo* tube)	To monitor the changes of endogenous metabolites in order to understand better the mechanism of action.	For 0–4 h after exposure: ↑ Uric acid, L-acetylcarnitine ↓ Azelaic acid, ophthalmic acid, suberic acid, ε-(γ-glutamyl)-lysine, pimelic acid For 12–24 h after exposure********: ↑ Ophthalmic acid, pimelic acid, suberic acid, azelaic acid, ↓ Uric acid, ε-(γ-glutamyl)-lysine, L-acetylcarnitine	• Application of DA-9701 affects purine metabolic pathway, lipid, fatty acid metabolism and lipid peroxidation. DA-9701 improves gastrointestinal motility.	[174]

Rats treated with AEE versus model.
**Patients from high busulfan concentration-time curve (high-AUC) compared with low-AUC group.*
***The differences between 2 weeks and 4 weeks of treatment.*
****Pre-treatment levels of metabolites compared to 24 hrs after the first infusion.*
*****Differences between responders and non-responders.*
******Type of association between post-treatment metabolites levels and post-treatment insulin measures.*
*******Fold change of mean urinary metabolite ratios in the induction phase.*
********When compared to mean fold-changes of 0–4 hours exposure.*

Table 3.
Examples of pharmacometabolomic studies.

is the evaluation of aspirin response variability during antiplatelet therapy [180]. Lewis and colleagues identified that metabolites related to aspirin (salicylic acid and 2-hydroxyhippuric acid) were significantly increased, but exposure to aspirin also changed the levels of purines, fatty acids, glycerol metabolites, amino acids, and carbohydrate-related metabolites. Moreover, a substantial difference could be observed between good and poor responders in purine metabolites - higher levels of inosine and adenosine were observed in poor responders after aspirin intervention. Later, the authors identified 51 SNPs in the *ADK* gene region that had associations with platelet aggregation in response to aspirin exposure, the strongest of which was the rs16931294 variant. To confirm their findings, the authors compared their results to previously obtained metabolomic data and observed that rs16931294 was significantly associated with adenosine monophosphate, xanthine, and hypoxanthine levels before aspirin exposure. When compared with post-exposure results, this SNP was strongly associated with levels of inosine and guanosine [180]. The examples presented above [142, 144, 146, 179, 180] as well as other available literature [36, 175, 178] demonstrate the importance of pharmacometabolomics in drug design studies. Combined with other approaches, *e.g.*, pharmacogenomic, pharmacometabolomics can greatly contribute to our understanding of individual differences in responses to drug treatment and thus directly aid us in the development of new generations of drugs. There is also potential for significantly extending our understanding of health sustenance and disease development, and thus reduce drug-dependent therapies. Perhaps not the most profitable news for the pharma industry, but good news for health workers and the general population who will be able to identify at risk individuals and indeed tailor health management strategies to prevent and/or reduce the impact of disease.

4. Future perspectives

An intense research on '-omics' approaches, devoted to human health, led to the development of pharmacometabolomics, which is a new horizon in personalized medicine. Numerous research data on metabolomics, genomics, and transcriptomics can be combined and compared with health records around the world due to potent databases and biobanks collecting data and samples. Nowadays, software and informatics systems with sophisticated algorithms of artificial intelligence allow for deeper analyses of pharmacometabolomics data, and transform general medicine into a personalized approach.

The analytical techniques, databases, and biobanks presented here are the general trends, which need to be further developed. The sensitivity of the analytical platforms needs to be improved, and additional ameliorations related to time and overall costs must be done. Particular attention must be paid to the standardization of study protocols. The number of data and samples deposited in databases and biobank must be extended.

Up to now, major efforts in pharmacometabolomics have been concentrated on research aspects and method validation for medical applications. The results presented here show undoubtedly that pharmacometabolomics is key for personalized medicine and needs to be transferred 'from bench to bedside'. Nevertheless, medical personnel can source from pharmacometabolomics only if the data are presented in a simple and comprehensive way. In the future, more effort is needed to increase the broad awareness of pharmacometabolomics among patients and healthcare system staff, and to introduce the benefits of pharmacometabolomics into clinical practice.

5. Conclusion

Human genetics and lifestyle variation directly influence pharmacological treatments, whose effect can be enhanced positively or negatively in some individuals over the statistical population used in clinical trials.

This chapter has described pharmacometabolomics as an innovative tool capable of assisting researchers and frontline medical personnel in establishing personalized therapeutic strategies. Pharmacometabolomics can be used to personalize treatment type, dosage, duration, and to monitor metabolites' profiles during pharmacotherapy. The existing '-omics' and health records databases, and biobanks of human fluid samples and tissues are a precious resource for pharmacometabolomics, which identify biomarkers of therapeutic effects over a disease course. The metabolomics databases are increasing their data pool every day, and are priceless for researchers combining '-omics' knowledge for better and personalized pharmacotherapy.

Acknowledgements

We would like to thank King Abdullah University of Science and Technology for financial support. We would like to acknowledge Life Science Editors for editorial services. Special thanks to Dr. Kristin Strandenes, EddaTxT, Norway for her suggestions and feedback.

Funding

We would like to thank King Abdullah University of Science and Technology (KAUST) for financial support.

Conflict of interest

The authors declare no conflict of interest.

Abbreviations

ADK	Adenosine kinase
ADR	Adverse drug reactions
AEE	Aspirin eugenol ester
ALI	Acute liver injury
ART	Antiretroviral therapy
BMRB	Biological Magnetic Resonance Data Bank
BRCA1	Breast cancer gene 1
BRCA2	Breast cancer gene 2
CDER	Center for Drug Evaluation and Research
CYP2D6	Cytochrome P450 2D6
DTX	Docetaxel
EHRs	Existing electronic health records
EMB	Ethambutol
ESI	Electro spray ionization
FDA	Food and Drug Administration

FT-IR	Fourier transformed infrared spectroscopy
HILIC	Hydrophilic interaction liquid chromatography
HMDB	Human Metabolome Database
HPLC	High-performance liquid chromatography
INH	Isoniazid
MATE1	Multidrug and toxin extrusion 1
MMCD	Madison Metabolomics Consortium Database
MS	Mass spectrometry
NMR	Nuclear magnetic resonance spectroscopy
OCT1	Organic cation transporter 1
OCT2	Organic cation transporter 2
PMRN	Pharmacometabolomics Research Network
PRIMe	Platform for RIKEN Metabolomics
PZA	Pyrazinamide
RIF	Rifampicin
RP	Reversed-phase gradient chromatography
SLC22A1	Solute carrier family 22 member 1
SLC22A2	Solute carrier family 22 member 2
SLC47A1	Solute carrier family 47, member 1
SNP	Single-nucleotide polymorphism
TMAO	Trimethylamine-N-oxide

Author details

Abdul-Hamid Emwas[1]*, Kacper Szczepski[2], Ryan T. McKay[3], Hiba Asfour[4], Chung-ke Chang[5], Joanna Lachowicz[6] and Mariusz Jaremko[2]

1 Core Labs, King Abdullah University of Science and Technology, Thuwal, Kingdom of Saudi Arabia

2 Biological and Environmental Sciences and Engineering Division (BESE), King Abdullah University of Science and Technology (KAUST), Thuwal, Saudi Arabia

3 Department of Chemistry, University of Alberta, Edmonton, Alberta, Canada

4 Birzeit University, West Bank, Birzeit, Palestine

5 Taiwan Biobank, Institute of Biomedical Sciences, Academia Sinica, Taiwan

6 Department of Medical Sciences and Public Health, Università di Cagliari, Cittadella Universitaria, Monserrato, Italy

*Address all correspondence to: abdelhamid.emwas@kaust.edu.sa

References

[1] Heindel JJ, McAllister KA, Worth JL, Tyson FL. Environmental Epigenomics, Imprinting and Disease Susceptibility. Epigenetics. 2006;1(1):2-7. DOI: 10.4161/epi.1.1.2642.

[2] Jirtle RL, Skinner MK. Environmental epigenomics and disease susceptibility. Nature Reviews Genetics. 2007;8(4):253-62. DOI: 10.1038/nrg2045.

[3] Emwas AH, Roy R, McKay RT, Tenori L, Saccenti E, Gowda GAN, Raftery D, Alahmari F, Jaremko L, Jaremko M, Wishart DS. NMR Spectroscopy for Metabolomics Research. Metabolites. 2019;9(7). DOI: 10.3390/metabo9070123.

[4] Eghbalnia HR, Romero PR, Westler WM, Baskaran K, Ulrich EL, Markley JL. Increasing rigor in NMR-based metabolomics through validated and open source tools. Current Opinion in Biotechnology. 2017;43:56-61. DOI: 10.1016/j.copbio.2016.08.005.

[5] Bruen D, Delaney C, Florea L, Diamond D. Glucose Sensing for Diabetes Monitoring: Recent Developments. Sensors. 2017;17(8): 1866. DOI: 10.3390/s17081866.

[6] Association AD. Classification and Diagnosis of Diabetes: Standards of Medical Care in Diabetes. Diabetes Care. 2020;43(Supplement 1):S14-S31. DOI: 10.2337/dc20-S002.

[7] Sherwani SI, Khan HA, Ekhzaimy A, Masood A, Sakharkar MK. Significance of HbA1c Test in Diagnosis and Prognosis of Diabetic Patients. Biomarker Insights. 2016;11:BMI. S38440. DOI: 10.4137/bmi.S38440.

[8] Chung L, Moore K, Phillips L, Boyle FM, Marsh DJ, Baxter RC. Novel serum protein biomarker panel revealed by mass spectrometry and its prognostic value in breast cancer. Breast Cancer Research. 2014;16(3):R63. DOI: 10.1186/bcr3676.

[9] Thelin E, Al Nimer F, Frostell A, Zetterberg H, Blennow K, Nyström H, Svensson M, Bellander B-M, Piehl F, Nelson DW. A Serum Protein Biomarker Panel Improves Outcome Prediction in Human Traumatic Brain Injury. Journal of Neurotrauma. 2019;36(20):2850-62. DOI: 10.1089/neu.2019.6375.

[10] Duangkumpha K, Stoll T, Phetcharaburanin J, Yongvanit P, Thanan R, Techasen A, Namwat N, Khuntikeo N, Chamadol N, Roytrakul S, Mulvenna J, Mohamed A, Shah AK, Hill MM, Loilome W. Discovery and Qualification of Serum Protein Biomarker Candidates for Cholangiocarcinoma Diagnosis. Journal of Proteome Research. 2019;18(9):3305-16. DOI: 10.1021/acs.jproteome.9b00242.

[11] Shimura T, Dayde D, Wang H, Okuda Y, Iwasaki H, Ebi M, Kitagawa M, Yamada T, Yamada T, Hanash SM, Taguchi A, Kataoka H. Novel urinary protein biomarker panel for early diagnosis of gastric cancer. British Journal of Cancer. 2020;123(11): 1656-64. DOI: 10.1038/s41416-020-01063-5.

[12] Guiraud S, Edwards B, Squire SE, Babbs A, Shah N, Berg A, Chen H, Davies KE. Identification of serum protein biomarkers for utrophin based DMD therapy. Scientific Reports. 2017;7(1):43697. DOI: 10.1038/srep43697.

[13] Liao H, Wu J, Kuhn E, Chin W, Chang B, Jones MD, O'Neil S, Clauser KR, Karl J, Hasler F, Roubenoff R, Zolg W, Guild BC. Use of mass spectrometry to identify protein biomarkers of disease severity in the synovial fluid and serum of patients with rheumatoid arthritis. Arthritis &

Rheumatism. 2004;50(12):3792-803. DOI: 10.1002/art.20720.

[14] Bansal N, Gupta AK, Gupta A, Sankhwar SN, Mahdi AA. Serum-based protein biomarkers of bladder cancer: A pre- and post-operative evaluation. Journal of Pharmaceutical and Biomedical Analysis. 2016;124:22-5. DOI: 10.1016/j.jpba.2016.02.026.

[15] Chai YD, Zhang L, Yang Y, Su T, Charugundla P, Ai J, Messadi D, Wong DT, Hu S. Discovery of potential serum protein biomarkers for lymph node metastasis in oral cancer. Head & Neck. 2016;38(1):118-25. DOI: 10.1002/hed.23870.

[16] Cheng Z, Yin J, Yuan H, Jin C, Zhang F, Wang Z, Liu X, Wu Y, Wang T, Xiao S. Blood-Derived Plasma Protein Biomarkers for Alzheimer's Disease in Han Chinese. Frontiers in Aging Neuroscience. 2018;10(414). DOI: 10.3389/fnagi.2018.00414.

[17] Sun J-L, Li S, Lu X, Feng J-B, Cai T-J, Tian M, Liu Q-J. Identification of the differentially expressed protein biomarkers in rat blood plasma in response to gamma irradiation. International Journal of Radiation Biology. 2020;96(6):748-58. DOI: 10.1080/09553002.2020.1739775.

[18] Kell DB. Systems biology, metabolic modelling and metabolomics in drug discovery and development. Drug Discovery Today. 2006;11(23):1085-92. DOI: 10.1016/j.drudis.2006.10.004.

[19] Urbanczyk-Wochniak E, Luedemann A, Kopka J, Selbig J, Roessner-Tunali U, Willmitzer L, Fernie AR. Parallel analysis of transcript and metabolic profiles: a new approach in systems biology. EMBO reports. 2003;4(10):989-93. DOI: 10.1038/sj.embor.embor944.

[20] Emwas A-HM, Al-Rifai N, Szczepski K, Alsuhaymi S, Rayyan S, Almahasheer H, Jaremko M, Brennan L, Lachowicz JI. You Are What You Eat: Application of Metabolomics Approaches to Advance Nutrition Research. Foods. 2021;10(6):1249.

[21] Vizán P, Mazurek S, Cascante M. Robust metabolic adaptation underlying tumor progression. Metabolomics. 2008;4(1):1-12. DOI: 10.1007/s11306-007-0101-3.

[22] Warburg O. On the Origin of Cancer Cells. Science. 1956;123(3191):309-14. DOI: 10.1126/science.123.3191.309

[23] Center for Drug Evaluation and Research. New drug therapy APPROVALS 2019 [Available from: https://www.fda.gov/drugs/new-drugs-fda-cders-new-molecular-entities-and-new-therapeutic-biological-products/new-drug-therapy-approvals-2019.

[24] FDA. Orange book: Approved drug products with therapeutic equivalence evaluations 2017 [Available from: https://www.fda.gov/drugs/drug-approvals-and-databases/approved-drug-products-therapeutic-equivalence-evaluations-orange-book.

[25] Knoben JE, Scott GR, Tonelli RJ. An overview of the FDA publication approved drug products with therapeutic equivalence evaluations. American journal of hospital pharmacy. 1990;47(12):2696-700. DOI: 10.1093/ajhp/47.12.2696.

[26] Bawa R, Bawa S, Mehra R. The translational challenge in medicine at the nanoscale. Audette, GF, Reese, BE, asst eds Handbook of Clinical Nanomedicine: Law, Business, Regulation, Safety and Risk, Pan Stanford Publishing, Singapore. 2016:1291-346. DOI: 10.1201/b19910

[27] Nair AK, Anand O, Chun N, Conner DP, Mehta MU, Nhu DT, Polli JE, Lawrence XY, Davit BM. Statistics on BCS classification of

generic drug products approved between 2000 and 2011 in the USA. The AAPS journal. 2012;14(4):664-6. DOI: 10.1208/s12248-012-9384-z.

[28] Nicholson JK, Wilson ID. Understanding 'Global' Systems Biology: Metabonomics and the Continuum of Metabolism. Nature Reviews Drug Discovery. 2003;2(8):668-76. DOI: 10.1038/nrd1157.

[29] Nicholson JK, Lindon JC, Holmes E. 'Metabonomics': understanding the metabolic responses of living systems to pathophysiological stimuli via multivariate statistical analysis of biological NMR spectroscopic data. Xenobiotica. 1999;29(11):1181-9. DOI: 10.1080/004982599238047.

[30] Bedair M, Sumner LW. Current and emerging mass-spectrometry technologies for metabolomics. Trac-Trends in Analytical Chemistry. 2008;27(3):238-50. DOI: 10.1016/j.trac.2008.01.006.

[31] Dudka I, Kossowska B, Senhadri H, Latajka R, Hajek J, Andrzejak R, Antonowicz-Juchniewicz J, Gancarz R. Metabonomic analysis of serum of workers occupationally exposed to arsenic, cadmium and lead for biomarker research: A preliminary study. Environment International. 2014;68:71-81. DOI: 10.1016/j.envint.2014.03.015.

[32] Cui GX, Liew YJ, Li Y, Kharbatia N, Zahran NI, Emwas AH, Eguiluz VM, Aranda M. Host-dependent nitrogen recycling as a mechanism of symbiont control in Aiptasia. Plos Genetics. 2019;15(6). DOI: 10.1371/journal.pgen.1008189.

[33] Guleria A, Pratap A, Dubey D, Rawat A, Chaurasia S, Sukesh E, Phatak S, Ajmani S, Kumar U, Khetrapal CL, Bacon P, Misra R, Kumar D. NMR based serum metabolomics reveals a distinctive signature in patients with Lupus Nephritis. Scientific Reports. 2016;6. DOI: 10.1038/srep35309.

[34] Huang Y, Tian Y, Li G, Li Y, Yin X, Peng C, Xu F, Zhang Z. Discovery of safety biomarkers for realgar in rat urine using UFLC-IT-TOF/MS and H-1 NMR based metabolomics. Analytical and Bioanalytical Chemistry. 2013;405(14):4811-22. DOI: 10.1007/s00216-013-6842-0.

[35] Stuart KA, Welsh K, Walker MC, Edrada-Ebel R. Metabolomic tools used in marine natural product drug discovery. Expert Opinion on Drug Discovery. 2020;15(4):499-522. DOI: 10.1080/17460441.2020.1722636.

[36] Mussap M, Loddo C, Fanni C, Fanos V. Metabolomics in pharmacology - a delve into the novel field of pharmacometabolomics. Expert Review of Clinical Pharmacology. 2020;13(2): 115-34. DOI: 10.1080/17512433.2020.1713750.

[37] Zhang S, Gowda GAN, Asiago V, Shanaiah N, Barbas C, Raftery D. Correlative and quantitative H-1 NMR-based metabolomics reveals specific metabolic pathway disturbances in diabetic rats. Analytical Biochemistry. 2008;383(1):76-84. DOI: 10.1016/j.ab.2008.07.041.

[38] Zhao Y, Fu L, Li R, Wang L-N, Yang Y, Liu N-N, Zhang C-M, Wang Y, Liu P, Tu B-B, Zhang X, Qiao J. Metabolic profiles characterizing different phenotypes of polycystic ovary syndrome: plasma metabolomics analysis. Bmc Medicine. 2012;10. DOI: 10.1186/1741-7015-10-153.

[39] Zheng H, Lorenzen JK, Astrup A, Larsen LH, Yde CC, Clausen MR, Bertram HC. Metabolic Effects of a 24-Week Energy-Restricted Intervention Combined with Low or High Dairy Intake in Overweight Women: An NMR-Based Metabolomics

Investigation. Nutrients. 2016;8(3). DOI: 10.3390/nu8030108.

[40] Dunn WB, Broadhurst DI, Atherton HJ, Goodacre R, Griffin JL. Systems level studies of mammalian metabolomes: the roles of mass spectrometry and nuclear magnetic resonance spectroscopy. Chemical society reviews. 2011;40(1):387-426. DOI: 10.1039/B906712B.

[41] Chandra K, Harthi S, Almulhim F, Emwas A-H, Jaremko L, Jaremko M. The robust NMR Toolbox for Metabolomics. Molecular Omics. 2021. DOI: 10.1039/D1MO00118C.

[42] Beltran A, Suarez M, Rodriguez MA, Vinaixa M, Samino S, Arola L, Correig X, Yanes O. Assessment of Compatibility between Extraction Methods for NMR- and LC/MS-Based Metabolomics. Analytical Chemistry. 2012;84(14):5838-44. DOI: 10.1021/ac3005567.

[43] Currie F, Broadhurst DI, Dunn WB, Sellick CA, Goodacre R. Metabolomics reveals the physiological response of Pseudomonas putida KT2440 (UWC1) after pharmaceutical exposure. Molecular Biosystems. 2016;12(4):1367-77. DOI: 10.1039/c5mb00889a.

[44] Lee BJ, Zhou YY, Lee JS, Shine BK, Seo JA, Lee D, Kim YS, Choi HK. Discrimination and prediction of the origin of Chinese and Korean soybeans using Fourier transform infrared spectrometry (FT-IR) with multivariate statistical analysis. Plos One. 2018;13(4). DOI: 10.1371/journal.pone.0196315.

[45] Qamar W, Ahamad SR, Ali R, Khan MR, Al-Ghadeer AR. Metabolomic analysis of lung epithelial secretions in rats: An investigation of bronchoalveolar lavage fluid by GC-MS and FT-IR. Experimental Lung Research. 2014;40(9):460-6. DOI: 10.3109/01902148.2014.947008.

[46] Al-Talla ZA, Akrawi SH, Tolley LT, Sioud SH, Zaater MF, Emwas AH. Bioequivalence assessment of two formulations of ibuprofen. Drug Des Devel Ther. 2011;5:427-33. DOI: 10.2147/DDDT.S24504 dddt-5-427 [pii].

[47] Liu M, Xie H, Ma Y, Li H, Li C, Chen L, Jiang B, Nian B, Guo T, Zhang Z, Jiao W, Liu Q, Ling T, Zhao M. High Performance Liquid Chromatography and Metabolomics Analysis of Tannase Metabolism of Gallic Acid and Gallates in Tea Leaves. Journal of Agricultural and Food Chemistry. 2020;68(17):4946-54. DOI: 10.1021/acs.jafc.0c00513.

[48] Chandra K, Al-Harthi S, Sukumaran S, Almulhim F, Emwas A-H, Atreya HS, Jaremko Ł, Jaremko M. NMR-based metabolomics with enhanced sensitivity. RSC Advances. 2021;11(15):8694-700. DOI: 10.1039/D1RA01103K.

[49] Emwas A-HM. The Strengths and Weaknesses of NMR Spectroscopy and Mass Spectrometry with Particular Focus on Metabolomics Research. In: Bjerrum JT, editor. Metabonomics: Methods and Protocols. New York, NY: Springer New York; 2015. p. 161-93 DOI: 10.1007/978-1-4939-2377-9_13.

[50] Wang JH, Byun J, Pennathur S. Analytical Approaches to Metabolomics and Applications to Systems Biology. Seminars in Nephrology. 2010;30(5):500-11. DOI: 10.1016/j.semnephrol.2010.07.007.

[51] Valentino G, Graziani V, D'Abrosca B, Pacifico S, Fiorentino A, Scognamiglio M. NMR-Based Plant Metabolomics in Nutraceutical Research: An Overview. Molecules. 2020;25(6):1444. DOI: 10.3390/molecules25061444.

[52] Zhang Y, Zhang H, Chang D, Guo F, Pan H, Yang Y. Metabolomics approach by 1H NMR spectroscopy of serum

reveals progression axes for asymptomatic hyperuricemia and gout. Arthritis Research & Therapy. 2018;20(1):111. DOI: 10.1186/s13075-018-1600-5.

[53] Emwas A-HM, Merzaban JS, Serrai H. Chapter 3 - Theory and Applications of NMR-Based Metabolomics in Human Disease Diagnosis. In: ur-Rahman A, Choudhary MI, editors. Applications of NMR Spectroscopy: Bentham Science Publishers; 2015. p. 93-130 DOI: 10.1016/B978-1-60805-963-8.50003-2.

[54] Emwas A-HM, Al-Talla ZA, Kharbatia NM. Sample Collection and Preparation of Biofluids and Extracts for Gas Chromatography–Mass Spectrometry. In: Bjerrum JT, editor. Metabonomics: Methods and Protocols. New York, NY: Springer New York; 2015. p. 75-90 DOI: 10.1007/978-1-4939-2377-9_7.

[55] Wang J, Li Y, Li S, Zhao W, Jiang R, Wang S. Application of mass spectrometry-based metabolomics in meat science: a review. Shipin Kexue / Food Science. 2020;41(23):293-302. DOI: 10.7506/spkx1002-6630-20200430-405.

[56] Do KT, Wahl S, Raffler J, Molnos S, Laimighofer M, Adamski J, Suhre K, Strauch K, Peters A, Gieger C, Langenberg C, Stewart ID, Theis FJ, Grallert H, Kastenmüller G, Krumsiek J. Characterization of missing values in untargeted MS-based metabolomics data and evaluation of missing data handling strategies. Metabolomics : Official journal of the Metabolomic Society. 2018;14(10):128-. DOI: 10.1007/s11306-018-1420-2.

[57] Zeng J, Wang Z, Huang X, Eckstein SS, Lin X, Piao H, Weigert C, Yin P, Lehmann R, Xu G. Comprehensive Profiling by Non-targeted Stable Isotope Tracing Capillary Electrophoresis-Mass Spectrometry: A New Tool Complementing Metabolomic Analyses of Polar Metabolites. Chemistry – A European Journal. 2019;25(21):5427-32. DOI: 10.1002/chem.201900539.

[58] Emwas A-HM, Al-Talla ZA, Kharbatia NM. Sample collection and preparation of biofluids and extracts for gas chromatography-mass spectrometry. Methods in molecular biology (Clifton, NJ). 2015;1277:75-90. DOI: 10.1007/978-1-4939-2377-9_7.

[59] Emwas A-HM, Al-Talla ZA, Yang Y, Kharbatia NM. Gas chromatography-mass spectrometry of biofluids and extracts. Methods in molecular biology (Clifton, NJ). 2015;1277:91-112. DOI: 10.1007/978-1-4939-2377-9_8.

[60] Fayek NM, Farag MA, Saber FR. Metabolome classification via GC/MS and UHPLC/MS of olive fruit varieties grown in Egypt reveal pickling process impact on their composition. Food Chemistry. 2021;339. DOI: 10.1016/j.foodchem.2020.127861.

[61] Gong YJ, Lyu WT, Shi XF, Zou XT, Lu LZ, Yang H, Xiao YP. A Serum Metabolic Profiling Analysis During the Formation of Fatty Liver in Landes Geese via GC-TOF/MS. Frontiers in Physiology. 2020;11. DOI: 10.3389/fphys.2020.581699.

[62] Loyo RM, Zarate E, Barbosa CS, Simoes-Barbosa A. Gas chromatography-mass spectrometry (GC/MS) reveals urine metabolites associated to light and heavy infections by *Schistosoma mansoni* in mice. Parasitology International. 2021;80. DOI: 10.1016/j.parint.2020.102239.

[63] Emwas AH, Luchinat C, Turano P, Tenori L, Roy R, Salek RM, Ryan D, Merzaban JS, Kaddurah-Daouk R, Zeri AC, Nagana Gowda GA, Raftery D, Wang Y, Brennan L, Wishart DS. Standardizing the experimental conditions for using urine in

NMR-based metabolomic studies with a particular focus on diagnostic studies: a review. Metabolomics. 2015;11(4):872-94. DOI: 10.1007/s11306-014-0746-7.

[64] Bhinderwala F, Wase N, DiRusso C, Powers R. Combining Mass Spectrometry and NMR Improves Metabolite Detection and Annotation. Journal of Proteome Research. 2018;17(11):4017-22. DOI: 10.1021/acs.jproteome.8b00567.

[65] Nageeb A, Al-Tawashi A, Mohammad Emwas A-H, Abdel-Halim Al-Talla Z, Al-Rifai N. Comparison of *Artemisia annua* Bioactivities between Traditional Medicine and Chemical Extracts. Current bioactive compounds. 2013;9(4):324-32. DOI: 10.2174/157340720904140404151439.

[66] Emwas AH, Szczepski K, Poulson BG, Chandra K, McKay RT, Dhahri M, Alahmari F, Jaremko L, Lachowicz JI, Jaremko M. NMR as a "Gold Standard" Method in Drug Design and Discovery. Molecules. 2020;25(20). DOI: 10.3390/molecules25204597.

[67] Naqi HA, Woodman TJ, Husbands SM, Blagbrough IS. 19 F and 1 H quantitative-NMR spectroscopic analysis of fluorinated third-generation synthetic cannabinoids. Analytical Methods. 2019;11(24):3090-100. DOI: 10.1039/C9AY00814D

[68] Markley JL, Brüschweiler R, Edison AS, Eghbalnia HR, Powers R, Raftery D, Wishart DS. The future of NMR-based metabolomics. Current Opinion in Biotechnology. 2017;43:34-40. DOI: 10.1016/j.copbio.2016.08.001.

[69] Emwas AH, Roy R, McKay RT, Ryan D, Brennan L, Tenori L, Luchinat C, Gao X, Zeri AC, Gowda GA, Raftery D, Steinbeck C, Salek RM, Wishart DS. Recommendations and Standardization of Biomarker Quantification Using NMR-Based Metabolomics with Particular Focus on Urinary Analysis. J Proteome Res. 2016;15(2):360-73. DOI: 10.1021/acs.jproteome.5b00885.

[70] Emwas AH, Saunders M, Ludwig C, Günther UL. Determinants for Optimal Enhancement in Ex Situ DNP Experiments. Applied Magnetic Resonance. 2008;34(3-4):483-94. DOI: 10.1007/s00723-008-0120-x.

[71] Qiu XH, Redwine D, Beshah K, Livazovic S, Canlas CG, Guinov A, Emwas AHM. Amide versus amine ratio in the discrimination layer of reverse osmosis membrane by solid state N-15 NMR and DNP NMR. Journal of Membrane Science. 2019;581:243-51. DOI: 10.1016/j.memsci.2019.03.037.

[72] Kovacs H, Moskau D, Spraul M. Cryogenically cooled probes—a leap in NMR technology. Progress in Nuclear Magnetic Resonance Spectroscopy. 2005;46(2-3):131-55. DOI: 10.1016/j.pnmrs.2005.03.001.

[73] Webb AG. Advances in Probe Design for Protein NMR. In: Webb GA, editor. Annual Reports on NMR Spectroscopy. 58: Academic Press; 2006. p. 1-50 DOI: 10.1016/S0066-4103(05)58001-3.

[74] Patra P, Bhanja SK, Sen IK, Nandi AK, Samanta S, Das D, Devi KSP, Maiti TK, Acharya K, Islam SS. Structural and immunological studies of hetero polysaccharide isolated from the alkaline extract of Tricholoma crassum (Berk.) Sacc. Carbohydrate Research. 2012;362:1-7. DOI: 10.1016/j.carres.2012.09.009.

[75] Shiomi N, Abe T, Kikuchi H, Aritsuka T, Takata Y, Fukushi E, Fukushi Y, Kawabata J, Ueno K, Onodera S. Structural analysis of novel kestose isomers isolated from sugar beet molasses. Carbohydrate Research. 2016;424:1-7. DOI: 10.1016/j.carres.2016.02.002.

[76] Blindauer CA, Emwas AH, Holý A, Dvořáková H, Sletten E, Sigel H. Complex Formation of the Antiviral 9-[2-(Phosphonomethoxy)Ethyl] Adenine (PMEA) and of Its N 1, N 3, and N 7 Deaza Derivatives with Copper(II) in Aqueous Solution. Chemistry – A European Journal. 1997;3(9):1526-36. DOI: 10.1002/chem.19970030922.

[77] Susapto HH, Alhattab D, Abdelrahman S, Khan Z, Alshehri S, Kahin K, Ge R, Moretti M, Emwas A-H, Hauser CAE. Ultrashort Peptide Bioinks Support Automated Printing of Large-Scale Constructs Assuring Long-Term Survival of Printed Tissue Constructs. Nano Letters. 2021;21(7):2719-29. DOI: 10.1021/acs.nanolett.0c04426.

[78] Emwas A-H, Alghrably M, Al-Harthi S, Poulson BG, Szczepski K, Chandra K, Jaremko M. New Advances in Fast Methods of 2D NMR Experiments. Nuclear Magnetic Resonance. 2019. DOI: 10.5772/intechopen.90263.

[79] Abdul Jameel AG, Alquaity ABS, Campuzano F, Emwas A-H, Saxena S, Sarathy SM, Roberts WL. Surrogate formulation and molecular characterization of sulfur species in vacuum residues using APPI and ESI FT-ICR mass spectrometry. Fuel. 2021;293:120471. DOI: https://doi.org/10.1016/j.fuel.2021.120471.

[80] Hajjar D, Kremb S, Sioud S, Emwas A-H, Voolstra CR, Ravasi T. Anti-cancer agents in Saudi Arabian herbals revealed by automated high-content imaging. PLoS ONE. 2017;12(6). DOI: 10.1371/journal.pone.0177316.

[81] Bouatra S, Aziat F, Mandal R, Guo AC, Wilson MR, Knox C, Bjorndahl TC, Krishnamurthy R, Saleem F, Liu P, Dame ZT, Poelzer J, Huynh J, Yallou FS, Psychogios N, Dong E, Bogumil R, Roehring C, Wishart DS. The human urine metabolome2013. e73076-e p. DOI: 10.1371/journal.pone.0073076.

[82] Zhou B, Xiao JF, Tuli L, Ressom HW. LC-MS-based metabolomics. Molecular bioSystems. 2012;8(2):470-81. DOI: 10.1039/c1mb05350g.

[83] Li F, Gonzalez FJ, Ma X. LC–MS-based metabolomics in profiling of drug metabolism and bioactivation. Acta Pharmaceutica Sinica B. 2012;2(2):118-25. DOI: 10.1016/j.apsb.2012.02.010.

[84] Liu J, Zhao M, Zhu Y, Wang X, Zheng L, Yin Y. LC–MS-Based Metabolomics and Lipidomics Study of High-Density-Lipoprotein-Modulated Glucose Metabolism with an apoA-I Knockout Mouse Model. Journal of Proteome Research. 2019;18(1):48-56. DOI: 10.1021/acs.jproteome.8b00290.

[85] Theodoridis GA, Gika HG, Want EJ, Wilson ID. Liquid chromatography-mass spectrometry based global metabolite profiling: A review. Analytica Chimica Acta. 2012;711:7-16. DOI: 10.1016/j.aca.2011.09.042.

[86] Theodoridis G, Gika HG, Wilson ID. LC-MS-based methodology for global metabolite profiling in metabonomics/metabolomics. Trac-Trends in Analytical Chemistry. 2008;27(3):251-60. DOI: 10.1016/j.trac.2008.01.008.

[87] Wang X, Li L. Mass Spectrometry for Metabolome Analysis. Mass Spectrometry Letters. 2020;11(2):17-24. DOI: 10.5478/MSL.2020.11.2.17.

[88] Yi L, Dong N, Yun Y, Deng B, Ren D, Liu S, Liang Y. Chemometric methods in data processing of mass spectrometry-based metabolomics: A review. Analytica Chimica Acta. 2016;914:17-34. DOI: 10.1016/j.aca.2016.02.001.

[89] Mohammed SAA, Khan RA, El-Readi MZ, Emwas AH, Sioud S, Poulson BG, Jaremko M, Eldeeb HM,

Al-Omar MS, Mohammed HA. Suaeda vermiculata Aqueous-Ethanolic Extract-Based Mitigation of CCl(4)-Induced Hepatotoxicity in Rats, and HepG-2 and HepG-2/ADR Cell-Lines-Based Cytotoxicity Evaluations. Plants (Basel). 2020;9(10). DOI: 10.3390/plants9101291.

[90] Dhahri M, Sioud S, Dridi R, Hassine M, Boughattas NA, Almulhim F, Al Talla Z, Jaremko M, Emwas A-HM. Extraction, Characterization, and Anticoagulant Activity of a Sulfated Polysaccharide from *Bursatella leachii* Viscera. ACS Omega. 2020;5(24):14786-95. DOI: 10.1021/acsomega.0c01724.

[91] Raji M, Amad M, Emwas AH. Dehydrodimerization of pterostilbene during electrospray ionization mass spectrometry. Rapid Commun Mass Spectrom. 2013;27(11):1260-6. DOI: 10.1002/rcm.6571.

[92] Bird IM. High performance liquid chromatography: principles and clinical applications. BMJ (Clinical research ed). 1989;299(6702):783-7. DOI: 10.1136/bmj.299.6702.783.

[93] Ingle KP, Deshmukh AG, Padole DA, Dudhare MS, Moharil MP, Khelurkar VC. Phytochemicals: Extraction methods, identification and detection of bioactive compounds from plant extracts. Journal of Pharmacognosy and Phytochemistry. 2017;6(1):32-6.

[94] Donato P, Cacciola F, Tranchida PQ, Dugo P, Mondello L. Mass spectrometry detection in comprehensive liquid chromatography: Basic concepts, instrumental aspects, applications and trends. Mass Spectrometry Reviews. 2012;31(5):523-59. DOI: 10.1002/mas.20353.

[95] Spagou K, Tsoukali H, Raikos N, Gika H, Wilson ID, Theodoridis G. Hydrophilic interaction chromatography coupled to MS for metabonomic/metabolomic studies. Journal of Separation Science. 2010;33(6-7):716-27. DOI: 10.1002/jssc.200900803.

[96] Appiah-Amponsah E, Owusu-Sarfo K, Gowda GAN, Ye T, Raftery D. Combining Hydrophilic Interaction Chromatography (HILIC) and Isotope Tagging for Off-Line LC-NMR Applications in Metabolite Analysis. Metabolites. 2013;3(3):575-91. DOI: 10.3390/metabo3030575.

[97] Fang Z-Z, Gonzalez FJ. LC-MS-based metabolomics: an update. Arch Toxicol. 2014;88(8):1491-502. DOI: 10.1007/s00204-014-1234-6.

[98] Sana TR, Waddell K, Fischer SM. A sample extraction and chromatographic strategy for increasing LC/MS detection coverage of the erythrocyte metabolome. Journal of Chromatography B-Analytical Technologies in the Biomedical and Life Sciences. 2008;871(2):314-21. DOI: 10.1016/j.jchromb.2008.04.030.

[99] Rodriguez-Morato J, Pozo OJ, Marcos J. Targeting human urinary metabolome by LC-MS/MS: a review. Bioanalysis. 2018;10(7):489-516. DOI: 10.4155/bio-2017-0285.

[100] Shimizu T, Watanabe M, Fernie AR, Tohge T. Targeted LC-MS Analysis for Plant Secondary Metabolites. In: António C, editor. Plant Metabolomics: Methods and Protocols. New York, NY: Springer New York; 2018. p. 171-81 DOI: 10.1007/978-1-4939-7819-9_12.

[101] Lu L, Wang J, Xu Y, Wang K, Hu Y, Tian R, Yang B, Lai Q, Li Y, Zhang W, Shao Z, Lam H, Qian P-Y. A High-Resolution LC-MS-Based Secondary Metabolite Fingerprint Database of Marine Bacteria. Scientific Reports. 2014;4(1):6537. DOI: 10.1038/srep06537.

[102] Penn E, Tracy DK. The drugs don't work? antidepressants and the current

and future pharmacological management of depression. Therapeutic Advances in Psychopharmacology. 2012;2(5):179-88. DOI: 10.1177/2045125312445469.

[103] Schork NJ. Personalized medicine: time for one-person trials. Nature News. 2015;520(7549):609. DOI: 10.1038/520609a.

[104] Karumanchi SA, Thadhani R. Kidney complications: Why don't statins always work? Nature Medicine. 2010;16(1):38-40. DOI: 10.1038/nm0110-38.

[105] Haro JM, Novick D, Bertsch J, Karagianis J, Dossenbach M, Jones PB. Cross-national clinical and functional remission rates: Worldwide Schizophrenia Outpatient Health Outcomes (W-SOHO) study. British Journal of Psychiatry. 2011;199(3):194-201. DOI: 10.1192/bjp.bp.110.082065.

[106] Piquette-Miller M, Grant DM. The Art and Science of Personalized Medicine. Clinical Pharmacology & Therapeutics. 2007;81(3):311-5. DOI: 10.1038/sj.clpt.6100130.

[107] Lazarou J, Pomeranz BH, Corey PN. Incidence of Adverse Drug Reactions in Hospitalized PatientsA Meta-analysis of Prospective Studies. JAMA. 1998;279(15):1200-5. DOI: 10.1001/jama.279.15.1200.

[108] Pokorska-Bocci A, Stewart A, Sagoo GS, Hall A, Kroese M, Burton H. 'Personalized medicine': what's in a name? Personalized Medicine. 2014;11(2):197-210. DOI: 10.2217/pme.13.107.

[109] Balashova EE, Maslov DL, Lokhov PG. A Metabolomics Approach to Pharmacotherapy Personalization. J Pers Med. 2018;8(3). DOI: 10.3390/jpm8030028.

[110] Li B, He X, Jia W, Li H. Novel Applications of Metabolomics in Personalized Medicine: A Mini-Review. Molecules. 2017;22(7). DOI: 10.3390/molecules22071173.

[111] Ziegelstein RC. Personomics: The Missing Link in the Evolution from Precision Medicine to Personalized Medicine. J Pers Med. 2017;7(4). DOI: 10.3390/jpm7040011.

[112] Roden DM, Altman RB, Benowitz NL, Flockhart DA, Giacomini KM, Johnson JA, Krauss RM, McLeod HL, Ratain MJ, Relling MV, Ring HZ, Shuldiner AR, Weinshilboum RM, Weiss ST. Pharmacogenomics: Challenges and Opportunities. Annals of Internal Medicine. 2006;145(10):749-57. DOI: 10.7326/0003-4819-145-10-200611210-00007.

[113] Eichelbaum M, Ingelman-Sundberg M, Evans WE. Pharmaco genomics and Individualized Drug Therapy. Annual Review of Medicine. 2006;57(1):119-37. DOI: 10.1146/annurev.med.56.082103.104724.

[114] Relling MV, Evans WE. Pharmacogenomics in the clinic. Nature. 2015;526(7573):343-50. DOI: 10.1038/nature15817.

[115] Malandrino N, Smith RJ. Personalized Medicine in Diabetes. Clinical Chemistry. 2011;57(2):231-40. DOI: 10.1373/clinchem.2010.156901.

[116] Kleinberger JW, Pollin TI. Personalized medicine in diabetes mellitus: current opportunities and future prospects. Annals of the New York Academy of Sciences. 2015;1346(1):45-56. DOI: 10.1111/nyas.12757.

[117] Pearson ER. Personalized medicine in diabetes: the role of 'omics' and biomarkers. Diabetic Medicine. 2016;33(6):712-7. DOI: 10.1111/dme.13075.

[118] Verma M. Personalized Medicine and Cancer. Journal of Personalized Medicine. 2012;2(1):1-14. DOI: 10.3390/jpm2010001.

[119] Gambardella V, Tarazona N, Cejalvo JM, Lombardi P, Huerta M, Roselló S, Fleitas T, Roda D, Cervantes A. Personalized Medicine: Recent Progress in Cancer Therapy. Cancers. 2020;12(4):1009. DOI: 10.3390/cancers12041009.

[120] Congiu T, Alghrably M, Emwas A-H, Jaremko L, Lachowicz JI, Piludu M, Piras M, Faa G, Pichiri G, Jaremko M, Coni P. Undercover Toxic Ménage à Trois of Amylin, Copper (II) and Metformin in Human Embryonic Kidney Cells. Pharmaceutics. 2021;13(6):830.

[121] Pearson ER, Donnelly LA, Kimber C, Whitley A, Doney AS, McCarthy MI, Hattersley AT, Morris AD, Palmer CN. Variation in TCF7L2 influences therapeutic response to sulfonylureas: a GoDARTs study. Diabetes. 2007;56(8):2178-82. DOI: 10.2337/db07-0440.

[122] Chan CWH, Law BMH, So WKW, Chow KM, Waye MMY. Novel Strategies on Personalized Medicine for Breast Cancer Treatment: An Update. International Journal of Molecular Sciences. 2017;18(11):2423. DOI: 10.3390/ijms18112423

[123] Amir E, Freedman OC, Seruga B, Evans DG. Assessing Women at High Risk of Breast Cancer: A Review of Risk Assessment Models. JNCI: Journal of the National Cancer Institute. 2010;102(10):680-91. DOI: 10.1093/jnci/djq088.

[124] Harbeck N, Penault-Llorca F, Cortes J, Gnant M, Houssami N, Poortmans P, Ruddy K, Tsang J, Cardoso F. Breast cancer. Nature Reviews Disease Primers. 2019;5(1):66. DOI: 10.1038/s41572-019-0111-2.

[125] Li Y, Steppi A, Zhou Y, Mao F, Miller PC, He MM, Zhao T, Sun Q, Zhang J. Tumoral expression of drug and xenobiotic metabolizing enzymes in breast cancer patients of different ethnicities with implications to personalized medicine. Scientific Reports. 2017;7(1):4747. DOI: 10.1038/s41598-017-04250-2.

[126] Olopade OI, Grushko TA, Nanda R, Huo D. Advances in Breast Cancer: Pathways to Personalized Medicine. Clinical Cancer Research. 2008;14(24):7988-99. DOI: 10.1158/1078-0432.Ccr-08-1211.

[127] Wishart DS, Knox C, Guo AC, Eisner R, Young N, Gautam B, Hau DD, Psychogios N, Dong E, Bouatra S, Mandal R, Sinelnikov I, Xia JG, Jia L, Cruz JA, Lim E, Sobsey CA, Shrivastava S, Huang P, Liu P, Fang L, Peng J, Fradette R, Cheng D, Tzur D, Clements M, Lewis A, De Souza A, Zuniga A, Dawe M, Xiong YP, Clive D, Greiner R, Nazyrova A, Shaykhutdinov R, Li L, Vogel HJ, Forsythe I. HMDB: a knowledgebase for the human metabolome. Nucleic Acids Research. 2009;37:D603-D10. DOI: 10.1093/nar/gkn810.

[128] Wishart DS, Tzur D, Knox C, Eisner R, Guo AC, Young N, Cheng D, Jewell K, Arndt D, Sawhney S, Fung C, Nikolai L, Lewis M, Coutouly M-A, Forsythe I, Tang P, Shrivastava S, Jeroncic K, Stothard P, Amegbey G, Block D, Hau DD, Wagner J, Miniaci J, Clements M, Gebremedhin M, Guo N, Zhang Y, Duggan GE, MacInnis GD, Weljie AM, Dowlatabadi R, Bamforth F, Clive D, Greiner R, Li L, Marrie T, Sykes BD, Vogel HJ, Querengesser L. HMDB: the human metabolome database. Nucleic Acids Research. 2007;35:D521-D6. DOI: 10.1093/nar/gkl923.

[129] Sakurai T, Yamada Y, Sawada Y, Matsuda F, Akiyama K, Shinozaki K, Hirai MY, Saito K. PRIMe Update:

Innovative Content for Plant Metabolomics and Integration of Gene Expression and Metabolite Accumulation. Plant and Cell Physiology. 2013;54(2):E5–+. DOI: 10.1093/pcp/pcs184.

[130] Ulrich EL, Akutsu H, Doreleijers JF, Harano Y, Ioannidis YE, Lin J, Livny M, Mading S, Maziuk D, Miller Z, Nakatani E, Schulte CF, Tolmie DE, Wenger RK, Yao HY, Markley JL. BioMagResBank. Nucleic Acids Research. 2008;36:D402-D8. DOI: 10.1093/nar/gkm957.

[131] Cui Q, Lewis IA, Hegeman AD, Anderson ME, Li J, Schulte CF, Westler WM, Eghbalnia HR, Sussman MR, Markley JL. Metabolite identification via the Madison Metabolomics Consortium Database. Nature Biotechnology. 2008;26(2):162-4. DOI: 10.1038/nbt0208-162.

[132] Wishart DS, Jewison T, Guo AC, Wilson M, Knox C, Liu YF, Djoumbou Y, Mandal R, Aziat F, Dong E, Bouatra S, Sinelnikov I, Arndt D, Xia JG, Liu P, Yallou F, Bjorndahl T, Perez-Pineiro R, Eisner R, Allen F, Neveu V, Greiner R, Scalbert A. HMDB 3.0-The Human Metabolome Database in 2013. Nucleic Acids Research. 2013;41(D1):D801-D7. DOI: 10.1093/nar/gks1065.

[133] Haug K, Salek RM, Conesa P, Hastings J, de Matos P, Rijnbeek M, Mahendraker T, Williams M, Neumann S, Rocca-Serra P, Maguire E, Gonzalez-Beltran A, Sansone S-A, Griffin JL, Steinbeck C. MetaboLights-an open-access general-purpose repository for metabolomics studies and associated meta-data. Nucleic Acids Research. 2013;41(D1):D781-D6. DOI: 10.1093/nar/gks1004.

[134] Salek, Haug K, Conesa P, Hastings J, Williams M, Mahendraker T, Maguire E, Gonzalez-Beltran AN, Rocca-Serra P, Sansone S-A, Steinbeck C. The MetaboLights repository: curation challenges in metabolomics. Database-the Journal of Biological Databases and Curation. 2013. DOI: 10.1093/database/bat029.

[135] Salek RM, Neumann S, Schober D, Hummel J, Billiau K, Kopka J, Correa E, Reijmers T, Rosato A, Tenori L, Turano P, Marin S, Deborde C, Jacob D, Rolin D, Dartigues B, Conesa P, Haug K, Rocca-Serra P, O'Hagan S, Hao J, van Vliet M, Sysi-Aho M, Ludwig C, Bouwman J, Cascante M, Ebbels T, Griffin JL, Moing A, Nikolski M, Oresic M, Sansone SA, Viant MR, Goodacre R, Gunther UL, Hankemeier T, Luchinat C, Walther D, Steinbeck C. COordination of Standards in MetabOlomicS (COSMOS): facilitating integrated metabolomics data access. Metabolomics. 2015;11(6):1587-97. DOI: 10.1007/s11306-015-0810-y.

[136] Allen NE, Sudlow C, Peakman T, Collins R. UK Biobank Data: Come and Get It. Science Translational Medicine. 2014;6(224). DOI: 10.1126/scitranslmed.3008601.

[137] Fan CT, Lin JC, Lee *C. Taiwan* Biobank: a project aiming to aid Taiwan's transition into a biomedical island. Pharmacogenomics. 2008;9(2):235-46. DOI: 10.2217/14622416.9.2.235.

[138] Kuriyama S, Yaegashi N, Nagami F, Arai T, Kawaguchi Y, Osumi N, Sakaida M, Suzuki Y, Nakayama K, Hashizume H, Tamiya G, Kawame H, Suzuki K, Hozawa A, Nakaya N, Kikuya M, Metoki H, Tsuji I, Fuse N, Kiyomoto H, Sugawara J, Tsuboi A, Egawa S, Ito K, Chida K, Ishii T, Tomita H, Taki Y, Minegishi N, Ishii N, Yasuda J, Igarashi K, Shimizu R, Nagasaki M, Koshiba S, Kinoshita K, Ogishima S, Takai-Igarashi T, Tominaga T, Tanabe O, Ohuchi N, Shimosegawa T, Kure S, Tanaka H, Ito S, Hitomi J, Tanno K, Nakamura M, Ogasawara K, Kobayashi S, Sakata K,

Satoh M, Shimizu A, Sasaki M, Endo R, Sobue K, Yamamoto M, Tohoku Med Megabank Project S. The Tohoku Medical Megabank Project: Design and Mission. Journal of Epidemiology. 2016;26(9):493-511. DOI: 10.2188/jea.JE20150268.

[139] Lin JC, Fan CT, Liao CC, Chen YS. Taiwan Biobank: making cross-database convergence possible in the Big Data era. Gigascience. 2017;7(1). DOI: 10.1093/gigascience/gix110.

[140] Lin J-C, Fan C-T, Liao C-C, Chen Y-S. Taiwan Biobank: making cross-database convergence possible in the Big Data era. GigaScience. 2017;7(1):1-4. DOI: 10.1093/gigascience/gix110.

[141] Watts G. UK Biobank opens its data vaults to researchers. BMJ : British Medical Journal. 2012;344. DOI: 10.1136/bmj.e2459.

[142] Clayton AT, Lindon JC, Cloarec O, Antti H, Charuel C, Hanton G, Provost J-P, Le Net J-L, Baker D, Walley RJ, Everett JR, Nicholson JK. Pharmaco-metabonomic phenotyping and personalized drug treatment. Nature. 2006;440(7087):1073-7. DOI: 10.1038/nature04648.

[143] Gowda GAN, Raftery D. Quantitating Metabolites in Protein Precipitated Serum Using NMR Spectroscopy. Analytical Chemistry. 2014;86(11):5433-40. DOI: 10.1021/ac5005103.

[144] Kaddurah-Daouk R, Baillie RA, Zhu H, Zeng Z-B, Wiest MM, Nguyen UT, Watkins SM, Krauss RM. Lipidomic analysis of variation in response to simvastatin in the Cholesterol and Pharmacogenetics Study. Metabolomics. 2010;6(2):191-201. DOI: 10.1007/s11306-010-0207-x.

[145] Leßig J, Fuchs B. Plasmalogens in Biological Systems: Their Role in Oxidative Processes in Biological Membranes, their Contribution to Pathological Processes and Aging and Plasmalogen Analysis. Current Medicinal Chemistry. 2009;16(16):2021-41. DOI: 10.2174/092986709788682164.

[146] Kaddurah-Daouk R, McEvoy J, Baillie RA, Lee D, Yao JK, Doraiswamy PM, Krishnan KRR. Metabolomic mapping of atypical antipsychotic effects in schizophrenia. Molecular Psychiatry. 2007;12(10):934-45. DOI: 10.1038/sj.mp.4002000.

[147] FDA-NIH Biomarker Working Group. Best (biomarkers, endpoints, and other tools): National Institutes of Health (US), Bethesda (MD); 2016.

[148] Farrell PM, Rosenstein BJ, White TB, Accurso FJ, Castellani C, Cutting GR, Durie PR, LeGrys VA, Massie J, Parad RB, Rock MJ, Campbell PW. Guidelines for Diagnosis of Cystic Fibrosis in Newborns through Older Adults: Cystic Fibrosis Foundation Consensus Report. The Journal of Pediatrics. 2008;153(2):S4-S14. DOI: 10.1016/j.jpeds.2008.05.005.

[149] Nagana Gowda GA, Raftery D. Quantitating metabolites in protein precipitated serum using NMR spectroscopy. Analytical chemistry. 2014;86(11):5433-40. DOI: 10.1021/ac5005103.

[150] Stone NJ, Robinson JG, Lichtenstein AH, Merz CNB, Blum CB, Eckel RH, Goldberg AC, Gordon D, Levy D, Lloyd-Jones DM, McBride P, Schwartz JS, Shero ST, Smith SC, Watson K, Wilson PWF. 2013 ACC/AHA Guideline on the Treatment of Blood Cholesterol to Reduce Atherosclerotic Cardiovascular Risk in Adults. Journal of the American College of Cardiology. 2014;63(25_Part_B):2889-934. DOI: doi:10.1016/j.jacc.2013.11.002.

[151] Ledermann J, Harter P, Matei D, Macpherson E, Watkins C, Carmichael J, Matulonis U, Gourley C, Friedlander M,

Vergote I, Rustin G, Scott C, Meier W, Shapira-Frommer R, Safra T. Olaparib Maintenance Therapy in Platinum-Sensitive Relapsed Ovarian Cancer. The New England journal of medicine. 2012;366(15):1382-92. DOI: 10.1056/NEJMoa1105535

[152] Basu NN, Ingham S, Hodson J, Lalloo F, Bulman M, Howell A, Evans DG. Risk of contralateral breast cancer in BRCA1 and BRCA2 mutation carriers: a 30-year semi-prospective analysis. Familial Cancer. 2015;14(4):531-8. DOI: 10.1007/s10689-015-9825-9.

[153] Senior JR. Evolution of the Food and Drug Administration Approach to Liver Safety Assessment for New Drugs: Current Status and Challenges. Drug Safety. 2014;37(1):9-17. DOI: 10.1007/s40264-014-0182-7.

[154] Struewing JP, Hartge P, Wacholder S, Baker SM, Berlin M, McAdams M, Timmerman MM, Brody LC, Tucker MA. The risk of cancer associated with specific mutations of BRCA1 and BRCA2 among Ashkenazi Jews. N Engl J Med. 1997;336(20):1401-8. DOI: 10.1056/nejm199705153362001.

[155] What is a context of use (COU) for a biomarker? 2018 [Available from: https://www.fda.gov/drugs/biomarker-qualification-program/context-use.

[156] Yerges-Armstrong LM, Ellero-Simatos S, Georgiades A, Zhu H, Lewis JP, Horenstein RB, Beitelshees AL, Dane A, Reijmers T, Hankemeier T, Fiehn O, Shuldiner AR, Kaddurah-Daouk R. Purine pathway implicated in mechanism of resistance to aspirin therapy: pharmacometabolomics-informed pharmacogenomics. Clin Pharmacol Ther. 2013;94(4):525-32. DOI: 10.1038/clpt.2013.119.

[157] Ellero-Simatos S, Lewis JP, Georgiades A, Yerges-Armstrong LM, Beitelshees AL, Horenstein RB, Dane A, Harms AC, Ramaker R, Vreeken RJ, Perry CG, Zhu H, Sanchez CL, Kuhn C, Ortel TL, Shuldiner AR, Hankemeier T, Kaddurah-Daouk R. Pharmacometabolomics reveals that serotonin is implicated in aspirin response variability. CPT Pharmacometrics Syst Pharmacol. 2014;3:e125. DOI: 10.1038/psp.2014.22.

[158] Ellero-Simatos S, Beitelshees AL, Lewis JP, Yerges-Armstrong LM, Georgiades A, Dane A, Harms AC, Strassburg K, Guled F, Hendriks MM, Horenstein RB, Shuldiner AR, Hankemeier T, Kaddurah-Daouk R, Pharmacometabolomics Research N. Oxylipid Profile of Low-Dose Aspirin Exposure: A Pharmacometabolomics Study. J Am Heart Assoc. 2015;4(10):e002203. DOI: 10.1161/JAHA.115.002203.

[159] Di Minno A, Porro B, Turnu L, Manega CM, Eligini S, Barbieri S, Chiesa M, Poggio P, Squellerio I, Anesi A, Fiorelli S, Caruso D, Veglia F, Cavalca V, Tremoli E. Untargeted Metabolomics to Go beyond the Canonical Effect of Acetylsalicylic Acid. J Clin Med. 2019;9(1). DOI: 10.3390/jcm9010051.

[160] Zhang ZD, Yang YJ, Liu XW, Qin Z, Li SH, Li JY. The Protective Effect of Aspirin Eugenol Ester on Paraquat-Induced Acute Liver Injury Rats. Front Med (Lausanne). 2020;7:589011. DOI: 10.3389/fmed.2020.589011.

[161] Ma N, Yang Y, Liu X, Li S, Qin Z, Li J. Plasma metabonomics and proteomics studies on the anti-thrombosis mechanism of aspirin eugenol ester in rat tail thrombosis model. J Proteomics. 2020;215:103631. DOI: 10.1016/j.jprot.2019.103631.

[162] Kim B, Lee JW, Hong KT, Yu KS, Jang IJ, Park KD, Shin HY, Ahn HS, Cho JY, Kang HJ.

Pharmacometabolomics for predicting variable busulfan exposure in paediatric haematopoietic stem cell transplantation patients. Sci Rep. 2017;7(1):1711. DOI: 10.1038/s41598-017-01861-7.

[163] Wu D, Li X, Zhang X, Han F, Lu X, Liu L, Zhang J, Dong M, Yang H, Li H. Pharmacometabolomics Identifies 3-Hydroxyadipic Acid, d-Galactose, Lysophosphatidylcholine (P-16:0), and Tetradecenoyl-l-Carnitine as Potential Predictive Indicators of Gemcitabine Efficacy in Pancreatic Cancer Patients. Frontiers in Oncology. 2020;9(1524). DOI: 10.3389/fonc.2019.01524.

[164] Combrink M, du Preez I, Ronacher K, Walzl G, Loots DT. Time-Dependent Changes in Urinary Metabolome Before and After Intensive Phase Tuberculosis Therapy: A Pharmacometabolomics Study. OMICS. 2019;23(11):560-72. DOI: 10.1089/omi.2019.0140.

[165] Sun Y, Kim JH, Vangipuram K, Hayes DF, Smith EML, Yeomans L, Henry NL, Stringer KA, Hertz DL. Pharmacometabolomics reveals a role for histidine, phenylalanine, and threonine in the development of paclitaxel-induced peripheral neuropathy. Breast Cancer Res Treat. 2018;171(3):657-66. DOI: 10.1007/s10549-018-4862-3.

[166] Chen L, Chen C-S, Sun Y, Henry NL, Stringer KA, Hertz DL. Feasibility of pharmacometabolomics to identify potential predictors of paclitaxel pharmacokinetic variability. Cancer Chemotherapy and Pharmacology. 2021. DOI: 10.1007/s00280-021-04300-7.

[167] Bao X, Wu J, Kim S, LoRusso P, Li J. Pharmacometabolomics Reveals Irinotecan Mechanism of Action in Cancer Patients. J Clin Pharmacol. 2019;59(1):20-34. DOI: 10.1002/jcph.1275.

[168] Bayet-Robert M, Morvan D, Chollet P, Barthomeuf C. Pharmacometabolomics of docetaxel-treated human MCF7 breast cancer cells provides evidence of varying cellular responses at high and low doses. Breast Cancer Res Treat. 2010;120(3):613-26. DOI: 10.1007/s10549-009-0430-1.

[169] Park JE, Jeong GH, Lee IK, Yoon YR, Liu KH, Gu N, Shin KH. A Pharmacometabolomic Approach to Predict Response to Metformin in Early-Phase Type 2 Diabetes Mellitus Patients. Molecules. 2018;23(7). DOI: 10.3390/molecules23071579.

[170] Elbadawi-Sidhu M, Baillie RA, Zhu H, Chen YI, Goodarzi MO, Rotter JI, Krauss RM, Fiehn O, Kaddurah-Daouk R. Pharmacometabolomic signature links simvastatin therapy and insulin resistance. Metabolomics. 2017;13. DOI: 10.1007/s11306-016-1141-3.

[171] Liu D, An Z, Li P, Chen Y, Zhang R, Liu L, He J, Abliz Z. A targeted neurotransmitter quantification and nontargeted metabolic profiling method for pharmacometabolomics analysis of olanzapine by using UPLC-HRMS. RSC Advances. 2020;10(31):18305-14. DOI: 10.1039/d0ra02406f.

[172] He C, Liu Y, Wang Y, Tang J, Tan Z, Li X, Chen Y, Huang Y, Chen X, Ouyang D, Zhou H, Peng J. (1)H NMR based pharmacometabolomics analysis of metabolic phenotype on predicting metabolism characteristics of losartan in healthy volunteers. J Chromatogr B Analyt Technol Biomed Life Sci. 2018;1095:15-23. DOI: 10.1016/j.jchromb.2018.07.016.

[173] Shin KH, Choi MH, Lim KS, Yu KS, Jang IJ, Cho JY. Evaluation of endogenous metabolic markers of

hepatic CYP3A activity using metabolic profiling and midazolam clearance. Clin Pharmacol Ther. 2013;94(5):601-9. DOI: 10.1038/clpt.2013.128.

[174] Jeong H-C, Park JE, Seo Y, Kim M-G, Shin K-H. Urinary Metabolomic Profiling after Administration of Corydalis Tuber and Pharbitis Seed Extract in Healthy Korean Volunteers. Pharmaceutics. 2021;13(4):522.

[175] Beger RD, Schmidt MA, Kaddurah-Daouk R. Current Concepts in Pharmacometabolomics, Biomarker Discovery, and Precision Medicine. Metabolites. 2020;10(4). DOI: 10.3390/metabo10040129.

[176] Kim J. Human Milk: The Original Personalized Medicine February 2016 [Available from: https://www.medela.us/breastfeeding-professionals/blog/human-milk-the-original-personalized-medicine.

[177] Verhasselt V. Breastfeeding, a personalized medicine with influence on short-and long-term immune health. Milk, Mucosal Immunity and the Microbiome: Impact on the Neonate. 2020;94:48-58. DOI: 10.1159/000505578.

[178] Kaddurah-Daouk R, Weinshilboum RM, Network PR. Pharmacometabolomics: Implications for Clinical Pharmacology and Systems Pharmacology. Clinical Pharmacology & Therapeutics. 2014;95(2):154-67. DOI: 10.1038/clpt.2013.217.

[179] Wikoff WR, Frye RF, Zhu H, Gong Y, Boyle S, Churchill E, Cooper-Dehoff RM, Beitelshees AL, Chapman AB, Fiehn O, Johnson JA, Kaddurah-Daouk R, Pharmacometabolomics Research N. Pharmacometabolomics Reveals Racial Differences in Response to Atenolol Treatment. PLOS ONE. 2013;8(3): e57639. DOI: 10.1371/journal.pone.0057639.

[180] Lewis JP, Yerges-Armstrong LM, Ellero-Simatos S, Georgiades A, Kaddurah-Daouk R, Hankemeier T. Integration of Pharmacometabolomic and Pharmacogenomic Approaches Reveals Novel Insights Into Antiplatelet Therapy. Clinical Pharmacology & Therapeutics. 2013;94(5):570-3. DOI: 10.1038/clpt.2013.153.

Chapter 4

Volatilomics of Natural Products: Whispers from Nature

Chiara Carazzone, Julie P.G. Rodríguez, Mabel Gonzalez and Gerson-Dirceu López

Abstract

Volatilomics studies the emission of volatile compounds from living organisms like plants, flowers, animals, fruits, and microorganisms, using metabolomics tools to characterize the analytes. This is a complex process that involves several steps like sample preparation, extraction, instrumental analysis, and data processing. In this chapter, we provide balanced coverage of the different theoretical and practical aspects of the study of the volatilome. Static and dynamic headspace techniques for volatile capture will be discussed. Then, the main techniques for volatilome profiling, separation, and detection will be addressed, emphasizing gas chromatographic separation, mass spectrometry detection, and non-separative techniques using mass spectrometry. Finally, the whole volatilome data pre-processing and multivariate statistics for data interpretation will be introduced. We hope that this chapter can provide the reader with an overview of the research process in the study of volatile organic compounds (VOCs) and serve as a guide in the development of future volatilomics studies.

Keywords: volatile organic compounds (VOCs), microbial volatile organic compounds (mVOCs), static headspace, dynamic headspace, SPME, PDMS-patches, GC–MS, metabolomics workflow

1. Introduction

Volatilomics indicates the qualitative and quantitative study of the volatilome, defined as the complex blend of volatile organic compounds (VOCs) originating from different biosynthetic pathways and emitted by living organisms [1]. VOCs are small molecules (below 500 Da), with hydrophobic character, low boiling points, and high vapor pressure at ambient temperature. Unconjugated volatiles can freely diffuse across membranes to be released from flowers, fruits, and vegetative tissues into the atmosphere and from roots into the soil to be perceived at short and long-distance. Therefore, plants and animals use VOCs for chemical communication with the surrounding ecosystem, and plants also use them as attractors for pollinators and defense against herbivory and biotic and abiotic stress [2–5].

The study of VOCs of plants has focused not only on the qualitative and quantitative composition of the volatile fraction but on the bioactive compounds as well as flavors and fragrances [6, 7]. Similarly, the understanding of fruits' sensorial attributes is of great interest as quality control, as well as in the determination of origin mark, and the performance of ecological studies aimed at the establishment of the relationship between the ripening stage and the incidence of fruit diseases for insect or microorganism attack [8–10].

Microorganisms produce a plethora of important microbial volatile organic compounds (mVOCs), that play an essential role in inter- and intra-kingdom connections. The study of mVOCs has allowed, for example, to detect terpenes, compounds normally associated with plants, also in fungi and bacteria [11]. Also, these compounds are related to ecological interactions between living organisms found in the soil, including the rhizosphere [12].

In addition, several studies of VOCs from animals not only have allowed decoding the signal of the animal chemical communication but also have demonstrated the potential use of that knowledge in early disease's diagnostics. For example, recent studies have shown novel practice for the detection of biomarkers to identify the intoxication using unusual biological fluids like ear wax, being fast, economic, and noninvasive bioanalysis, with minimal sample preparation and very versatile to identify the first signals of intoxication [13, 14].

Differently to the genomes, the volatilome changes continuously across time, and its composition depends on external and internal factors, such as the environmental conditions, and/or the physiological state [15]. Therefore, the study of the volatilome is not a simple task and the researchers in this area entail multiple challenges derived from the chemical complexity of the samples and the superposition of VOCs signals as proper of the ecosystems. Thus, sensitive yet unbiased methodologies are needed to provide researchers with comprehensive and accurate representations of a plant species' volatile metabolome.

However, current methodologies are limited in their ability to isolate, and even more critically to identify, many of the compounds present in each sample. In volatile metabolomics, the emitted metabolites are already isolated from tissues, they need to be temporarily trapped, and eventually preconcentrated, in a way that allows them to be released unadulterated for separation and identification.

A variety of technologies have been developed. In these methods, the sample of interest is enclosed in a collection chamber and the released volatiles present in the airspace surrounding the sample, headspace (HS), are trapped onto an adsorbent. And are subsequently analyzed by gas chromatography in combination with mass spectrometry (GC–MS) as the method of choice for volatilomics.

Hence, in the next sections of this chapter, we will provide an overview of the volatilome study process, including the main practical and theoretical aspects of volatiles capture, sample preparation, and the main analytical techniques employed to monitor VOCs, together with the chemoinformatics tools used for volatilome dereplication, elucidation, annotation, and interpretation of data.

2. Volatiles collection

Sample acquisition in volatilomics experiments requires consistency, therefore due to the high variability of chemical structures, concentrations levels, sample types, and physiological variations, other variables different than metabolites (addressed as meta-variables from now on) should be controlled or at least carefully monitored in order to evaluate their effect on the study outcome. Some important variables that should be taken into account include replicate number, taxonomic identification, geographic location, phenotypic or phylogenetic variant, sample weight, phenotypic characteristics, sex, developmental stage, health status, collection date, and time. Photographs should be taken. A useful reference for registering meta-variables is the ReDU Sample Information Template. (https://docs.google.com/spreadsheets/d/1v71bnUd8fiXX51zuZIUAvYETWmpwFQj-M3mu4CNsHBU/

edit?usp=sharing) [16] build by the collaborative Global Natural Products Social Networking (GNPS) (https://gnps.ucsd.edu/ProteoSAFe/static/gnps-splash.jsp) [17] where researchers can add new meta-variables and share their data in an open-source and collaborative environment.

2.1 Plants

The plant volatilome is defined as the complex blend of essential oils (EOs) and VOCs fed by different biosynthetic pathways and emitted by plants, constitutively and/or after induction, as a defense strategy against biotic and abiotic stress. Plants have a vast diversity in their range of metabolites and their concentrations, as there are hundreds of thousands of metabolites in different categories. As such, there is no single analytical technique that has the capability of extracting and detecting the whole metabolome [18].

Plant volatile emissions are linked to the physiological status of the emitter, therefore special care must be taken to control the plant-growing environment as well as all variables concerning the developmental stage of the plant to limit unwanted fluctuations in metabolism that might affect collected. These include the time of day, photoperiod, temperature, humidity, water conditions, collection site altitude, plant age, climate, and soil type so that a careful experimental design is recommended. Whenever possible, growth chambers must be used for plant cultivation and volatile collection [19, 20]. EOs and VOCs can be extracted and analyzed from both fresh and dried plant materials. When using fresh material, particular attention must be paid to the health status of plants, since microbial and other infections may alter metabolites production. Plants must not show necrotic areas and be at the same developmental stage if comparative analyses are needed. Since the content of water may vary, it is a good practice to use some of the fresh material to calculate the dry matter percentage [21].

Since volatile emissions from many plant species vary with respect to the time of day, and different organs in the plant are known to produce and/or accumulate different profiles of secondary metabolites, collection strategies should consider volatile sampling over an extended period of time and from the investigated organ or entire plant, to prevent unintentional exclusion of volatile components in the sampled mixture. Also, when running VOCs analyses from living plants it must be remembered that rooted plants in pots respond differently than cuttings, and that soil in pots may contain microorganisms that can produce VOCs [22, 23]. Once a plant part is collected, at least two herbarium samples should be prepared and identified or authenticated by a taxonomist. One of these voucher specimens should be deposited in a local national herbarium. A card with details of the place, altitude, environment, and photographs should be attached to the herbarium sample, in case a recollection of the plant material is necessary. Although depositing herbarium samples is a basic step in performing phytochemical investigation, many researchers in the past neglected this step and thus were unable to reproduce their work [23–25].

2.2 Flowers

Living flowers change their volatile profile in a continuous way that depends on intrinsic and extrinsic factors. Once cut, flowers undergo rapid deterioration and loose volatiles. Flower volatiles allow discrimination between different plants and attract insects for pollination when they are released. The amount of emission is not uniform through time, with some differences between diurnal and nocturnal emission levels, and between reproduction phases. The volatile compounds emitted by flowers are mainly aliphatics, terpenoids, benzenoids, and phenylpropanoids. Flower volatiles require special methods for their isolation with preconcentration and can be obtained from the air surrounding the living or excised flower, or from the flower tissues themselves. The selected extraction technique determines the composition of the isolated volatiles mixture [26, 27].

2.3 Fruits

Fruits are very complex samples, rich in a great number of different classes of metabolites, including volatile, semi-volatile, and no volatile compounds. The flavor is one of the most important characteristics to value the quality of fruit. Volatile and semi-volatile compounds usually are responsible for aroma fruit, and their study has conducive to identify both positive and negative sensory attributes [28]. VOCs are produced in trace amounts, and although they are easily perceptible by the human nose, their sampling and monitoring can be challenging at an analytical level [29]. The volatile fraction of fruits is composed of hundreds of different chemical substances that can vary according to the type of fruit, but the emitted compounds can be grouped according to the chemical function mainly into esters, alcohols, aldehydes, ketones, lactones, and terpenoids [29]. Moreover, VOCs emitted by fruit depend on the production conditions (cultivars, state of maturity, post-harvest treatment, and storage) the sample format (whole fruit, sliced, wet, dry), and the type of analysis (in-field or in-lab). Capturing volatiles in-situ is a challenge, as small amounts of VOCs are released and diffuse in a large volume of air, which requires highly efficient sampling techniques to capture them. Solid-phase microextraction (SPME) and solid-phase extraction (SPE) are usually the most profitable techniques for the capture of fruit volatiles in-situ. Once the volatile compounds are retained in an adsorbent material, their storage and transport are facilitated. On the other hand, in laboratory capture of VOCs from fruits, can be efficiently performed by solvent or gas-based extraction techniques, such as Soxhlet, simultaneous distillation extraction, purge and trap, and headspace, among others.

2.4 Microorganisms

Analysis of mVOCs is commonly performed under controlled culture media, temperature, and agitation. Also, the percentage of humidity and exposure to UV–visible light among other growing conditions should be taken into account. In order to account for reproducibility of the experiments, laboratory tests on microorganisms must be performed using international reference strains e.g.: American Type Culture Collection (ATCC), instead of clinical or field isolations, or even strains isolated and saved in the research group for a long time. Because the emission of VOCs can vary in terms of presence or absence, and in terms of fluctuation in concentration, throughout the life span of the microorganisms (which can be from a few hours to days), it is advisable to perform analyses both in the exponential or logarithmic growth phase, as well as in the stationary phase [12, 30, 31]. During the exponential phase, the microorganism is reactivating its biosynthetic pathways after having been in a state of latency. Therefore, in this stage, there is generally a high concentration of some metabolites that are part of the first stages of the biosynthetic pathways, which can later diminish and disappear in the exponential phase. The stationary phase is achieved when the initial metabolic processes have been reached and occurs when the survival process of the species begins [32]. The metabolic changes produced in these two stages of microbial culture are fundamental to understanding and solving research questions [33, 34]. The determination of each of the culture phases is commonly done with a measurement of the absorption of light in the visible region between 500 and 650 nm for liquid growth medium. This is achieved by counting the colony-forming units (CFU) in the solid medium. The sampling time for analysis of mVOCs must coincide with those obtained in the growth curves, correctly differentiating the exponential and stationary phases.

2.5 Animals

For conducting volatile sampling from animals, the specimens could be either raised in captivity at controlled vivaria or extracted from their natural environments. Proper training in animal manipulation is an important aspect to be fulfilled before performing animal experimentation, as well as an approved permit by the Institution in charge to validates the procedures. Also, when animals are to be collected in their habitats, it is necessary to review if a Convention on International Trade in Endangered Species of Wild Fauna and Flora (CITES) permit is needed for protected species. A specialist should validate taxonomic identification and, in those cases, where sample collection involves euthanization of specimens that should be registered at a recognized Museum, and voucher numbers should be annotated and published on the research paper. In the same way, as other organisms could be sampled by different methods, almost all animals could be sampled in vivo, but in some cases, tissue extraction could be preferred for guaranteeing detection of less abundant metabolites. Some techniques applied for VOCs analysis from terrestrial arthropods [35–39], aquatic organisms [40–42], mammals [43–48], birds [49, 50], reptiles [51, 52], fishes [53], and amphibians [37, 38, 54–57] include headspace-adsorbent traps, polydimethylsiloxane (PDMS) patches, swabs and stir bar sorptive extraction (SBSE).

3. Volatiles extraction

Sample preparation is one of the most important steps in the analytical process. The goal of sample preparation is to efficiently isolate target analytes from potential interferences and to extract as many VOCs as possible to provide a true representation of the studied system.

Some steps of pre-treatment of the sample are necessary in order to minimize the manipulation of the sample and avoid its modification, to clean-up the sample efficiently, and to quench metabolic reactions that could cause degradation and decomposition. To date, two different types of headspace sampling, static and dynamic, are widely used for volatilomics investigation.

Static headspace sampling is a passive technique for VOCs collection, where no air is circulated for the concentration of the volatiles on a sorbent matrix [18]. As a result, the background noise is drastically reduced due to the absence of a continuous airflow that can contain impurities that could mask compounds released at trace amounts. In static headspace methods, samples are typically sealed inside a container or bag, where the volatiles are released and, in the more traditional version of the technique, the headspace is sampled directly using a gas-tight syringe and transferred to the Gas Chromatography (GC) injection port. When the analytes are present at trace level, it might be necessary to carry out static headspace methods with special techniques to concentrate volatiles during collection and reduce the dilution of the sample during desorption in the GC inlet. In such a context, SPME stands out as the most versatile strategy for volatile capture from the sample headspace in static mode. Nowadays, SPME is the leading technique in the analysis of volatiles of biological origin because it uses a fiber coated with a sorbent phase to combine extraction and pre-concentration compounds. SPME fibers are available in a wide range of coatings that allow the sampling of compounds of different polarities and volatilities. Considering that the goal of volatilomics profiling is to analyze as many metabolites as possible, the use of divinylbenzene/carboxen/polydimethylsiloxane (DVB/CAR/PDMS) fibers is the most suited to increase the number of analytes that can be trapped on the fiber because it can allow capture VOCs in a wide range of polarity and molecular weight [58].

This type of coating contains a layer of CAR particles underneath a layer of DVB particles. Because the ability of adsorbent coatings to extract a particular analyte strongly depends on the size of the pores, larger analytes will be retained in the outer DVB layer, while the smaller analytes will migrate through this layer and are retained by the inner layer of CAR. On the contrary, if the study targeted only on the most volatile fraction, PDMS/CAR would be an appropriate choice of coating, since the micropores of the CAR retain smaller analytes better than other coatings, although introducing a high degree of discrimination towards high-molecular-weight compounds.

On another hand, although other coatings, such as PDMS, polyacrylate (PA), and Carbowax (CW), are also commercially available, their use in volatilomics is quite scarce due to the higher selectivity towards certain classes of polarities [58, 59].

From a practical point of view, SPME is a versatile technique for in-field sampling as a non-destructive strategy for the study of the volatiles emitted ex-vivo, for example, by grapes. In this case, an aluminum wire cage can be used to support a polymeric film to enclose a whole cluster of grapes, and SPME fiber is introduced through a port fitted with a silicone septum (**Figure 1a**).

Also, an interesting strategy for speeding up the volatiles' uptake is vacuum-assisted SPME. For example, in-field sampling of volatiles from a single grape berry, a modified screw top, and a 2 mL glass vial can be used for fiber exposition. A syringe is usually used to create a negative pressure to hold the sampling device with the SPME sealed onto the sample surface (**Figure 1b**).

This type of coating contains a layer of CAR particles underneath a layer of DVB particles. Because the ability of adsorbent coatings to extract a particular analyte strongly depends on the size of the pores, larger analytes will be retained in the outer DVB layer, while the smaller analytes will migrate through this layer and are retained by the inner layer of Carboxen. Conversely, if the study targeted only on the most volatile fraction, PDMS/CAR would be an appropriate choice of coating, since the micropores of the CAR retain smaller analytes better than other coatings, although introducing a high degree of discrimination towards high-molecular-weight compounds.

Although SPME generally exhibits better extraction efficiency as the polarity of the compound decreases, these three coatings can provide balanced metabolome

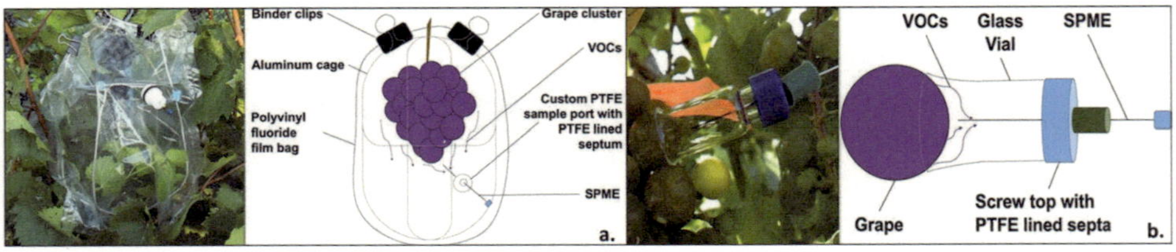

Figure 1.
Sampling handling techniques of VOC's fruit. a) Ex-vivo sampling of volatiles from the whole cluster of grapes by SPME; b) Ex-vivo sampling of volatiles from a single grape berry by SPME. Adapted from [10] under the Creative Commons Attribution License.

coverage as long as most polar analytes are present at reasonable concentration levels. Absorbent coatings, such as PDMS, PA, and CW, were rarely employed in profiling studies. These coatings display selectivity based on polarity, resulting in poor metabolomic coverage. The second case is dynamic headspace sampling, which offers a highly concentrated sample that can be desorbed into a solvent at volumes suitable for multiple analyses. To date, it is the most frequently used technique in all areas of plant volatile analysis. Dynamic headspace sampling collects a much larger quantity of compounds at higher concentrations because the continuous stream of air allows the sorbent to act as a filter trapping the volatiles.

Also, push and pull headspace sampling, two examples of dynamic headspace sampling, allow to avoid problems often encountered with the sealed systems used in static headspace and closed-loop stripping methods including heat, water vapor, and, in the case of plants, ethylene accumulation that can affect not only sampling efficiency but also plant physiology. Among the several methods, closed-loop stripping systems have broad utility for the collection of volatiles: volatiles are collected during continuous circulation of HS air inside closed chambers in which air circulation pumps are connected to supporting columns or coated supports [22].

As an example, SPME is a versatile technique for in-field sampling handling as a non-destructive strategy for the study of the volatiles emitted ex-vivo, for example, by the whole cluster of grapes. In this case, an aluminum wire cage can be used to support a polymeric film to enclosing a whole cluster of grapes, and SPME fiber introduced through a port fitted with silicone septa (**Figure 1a**) [60]. Also, an interesting strategy for speed up the volatile's uptake is vacuum-assisted SPME. For example, in-field sampling of volatiles from a single grape berry, a modified screw top, and a 2 mL glass vial can be used for fiber exposition. A syringe is usually used to create a negative pressure to hold the sampling device with SPME sealed unto the sample surface (**Figure 1b**) [60].

Alternatively, to the SPME, some liquid-phase microextractions (LPME), such as the single drop microextraction (SDME) or the hollow fiber liquid-phase microextraction (HF-LPME), can also provide efficient and profitable volatiles recoveries in the headspace static mode. For example, SDME is a technique based on a few microliters of solvent, in which volatiles can be capture in a small drop of extraction solvent-exposed to the headspace of the sample [20, 59]. In the same way, to address the drawbacks of the drop instability, the extraction solvent can be deposited into the lumen of a porous fiber HF-LPME, improving the extraction kinetics by use of a bigger transference surface or by the incorporation of an acceptor solvent into the membrane pores (Supported Liquid Membrane, SLM). Although the use of hazardous organic solvents can be considered a drawback, nowadays those solvent-based extractions can be performed with environmental-friendly alternatives, such as ionic liquids, deep eutectic solvents, or supramolecular solvents, among others.

The second type of headspace sampling is the dynamic headspace (DHS) method. It encompasses strategies in which VOCs are captured in a sorbent-packed trap by passing a continuous flow of inert dry gas through the sample. In this way, the emission of VOCs speeds up by the continuous renovation of the headspace fraction. After extraction, concentrated VOCs can be desorbed from the sorbent-packed trap with a suitable solvent or via thermal desorption. Besides, DHS address some drawbacks of the static modes such as the accumulation of water vapor or highly concentrated compounds, which presence can affect extraction efficiency. Two examples of dynamic headspace sampling which allow avoiding some drawbacks of the static mode, e.g., heat and water vapor accumulation that can affect not only sampling efficiency but also plant physiology, are closed-loop stripping and push and pull methods. These systems collect VOCs in sorbent-packed traps or coated devices, via the continuous circulation of gas inside closed circuits [22].

In addition to headspace sampling techniques, some sui generis approaches can combine two methods from different groups, for example, solvent-assisted flavor evaporation (SAFE). SAFE is an exhaustive extraction technique based on the high volatility rather than the polarity of the target compounds. In this case, a crude-extract from dry sample pieces is prepared with an appropriate solvent, such as dichloromethane, and then added into the dropping funnel and passed through a specific distillation chamber. Extraction takes place at high vacuum, and low-temperature conditions (20–30°C), and VOCs are collected in a cooled extraction vessel [61]. Other techniques including in this group are simultaneous extraction-distillation (SDE) and/or liquid–liquid extraction (LLE). Nevertheless, those can be subjected to some drawbacks, like the use of hazardous solvents, as well as the requirement of high temperatures and long extraction times, with potential formation of artifacts and degradation of some compounds.

Finally, volatile compounds also can be obtained for direct collection of the secretions of odoriferous glands or via non-invasive strategies using PDMS patches or swabs [22]. These techniques are especially useful in the monitoring of VOCs from animals. For example, obtaining the animal skin volatilome on PDMS patches is an excellent option [62]. Patches could be prepared by cutting a Silicone Elastomer Sheet (Goodfellows mfr. No. 942-965-49, Coraopolis, PA) and then carefully fix it on the animal skin with Tegaderm® dressings or water block clear Band-aids®. Alternatively, this procedure could be modified by gently swabbing the skin with or without previous stress-induced secretion. PDMS patches also can be placed into an animal enclosure and used without direct contact for capturing the volatiles that emanates in the headspace.

4. Volatilome profiling: separation and detection

Currently, gas chromatography coupled to mass spectrometry (GC–MS) is the primary analytical technique for the elucidation of the volatilome profile from natural sources. In gas chromatography analytes elute according to their volatility carried by a gas, usually Helium, through a coated fused silica capillary using a temperature gradient. Separation occurs based on the differential partition between the gas phase and the coating and the eluting peaks will give a response in the detector. The sample is vaporized in the injection system before it enters the column.

Several injection systems can be used to introduce the sample onto the column. Split injection allows transferring to the column only controlled sample amounts and prevent overloading of the column, thanks to a split valve at the base of the hot injector that divides the flow between column and waste in a fixable ratio. High-concentration samples can easily overload the GC column, resulting in all active sites on the column becoming occupied and leading to additional analytes not being retained and therefore to poor chromatographic resolution. For trace analysis, the injector can be used in splitless mode, which allows the entire volume of sample vaporized in the injector to reach the column. An alternative to the split/splitless interface is the programmed temperature vaporizer (PTV). Samples are injected onto a cool (40–60°C) PTV where they are trapped and concentrated on different sorbent materials before the inlet is rapidly heated to desorb the sample onto the column.

Different selectivity and sizes of columns have been used for GC–MS–based metabolomic analysis. The most used phase is 5% phenyl, 95% methyl siloxane, which offers a sufficiently generic selectivity, optimal for metabolomic applications where analytes with a wide range of volatilities have to be separated. Capillary columns of 25 to 30 m will provide the highest resolution and are available in most phases. An important point for all capillary GC–MS work is the need to condition the column prior to running valuable samples. Sangster et al. have recommended that several quality control samples be run at the beginning of a sample batch to condition the column [63]. Care also needs to be taken to randomize the injection sequence in order not to compromise subsequent statistical analysis.

In GC–MS ionization of analytes is mainly produced by electron ionization (EI) or chemical ionization (CI), while ion separation is obtained by mass analyzers operating on different principles. In EI, analytes that elute from the GC column are vaporized into the ion source and collide with an electron beam at 70 eV. As a result of the high energy imparted by electrons to the vaporized molecules, characteristic fragmentation occurs, providing structural information. EI is very robust and highly reproducible between instruments, and spectral libraries are available that can be used to search for the identities of unknown compounds based on m/z and intensity ratios of the observed fragment ions. A disadvantage of EI is that fragmentation is usually so efficient that the intensity of the molecular ion can be extremely low or even lost. For CI, a reagent gas, such as methane or ammonia, is introduced into the source of the mass spectrometer. Protonated gas ions, produced by the collision with electrons originating from an electron beam, ionize the analytes eluting from the column after vaporization into the ion source. Significantly less energy than in EI is transferred to the analytes, and as a result, the dominant ion is usually the molecular ion.

Mass spectrometer based detectors are mainly used in metabolomic analysis and can be grouped according to the spectral information they provide, i.e., low-resolution instruments such as quadrupole mass spectrometer (qMS), ion-trap mass spectrometer (IT-MS), and high-speed time-of-flight mass spectrometer (TOF-MS) give nominal molecular weights and fragmentation of an analyte, while high-resolution instruments (high-resolution TOF-MS and hybrids) give the precise elemental composition of nominal masses. The single quadrupole mass analyzer is widely used and relatively inexpensive. The ions move along the axis of four parallel rods to which a direct current (DC) and an alternating current (AC) voltage are applied. These voltages affect the trajectory of ions traveling down the flight path between the rods in a way that only ions of a given m/z are transmitted at a given point in time. Scan speeds are rather low on quadrupole instruments, therefore considering the very high separation power of GC with peak widths of only a few seconds, it will be difficult to acquire several spectra across the width of a typical peak on a single quadrupole instrument. Time-of-flight (TOF) instruments are the most common mass analyzers in GC–MS–based metabolomics. The ions are accelerated in an electric field in which ions with the same charge will have the same kinetic energy, but different velocity depending on their mass-to-charge ratio (m/z). Successively, the ions enter a field-free region (flight tube) where they separate based on their m/z. TOF instruments are characterized by the fastest scan rate among all mass analyzers: a significant number of spectra can be acquired across each peak, leading to higher sensitivity and better spectral quality.

GC–MS has very high sensitivity and can therefore be used for the analysis of less commonly encountered samples that might only be available in trace amounts. Monodimensional GC–MS analysis provides suitable resolving-power for the analysis of relatively simple mixtures of VOCs. Nevertheless, volatilome samples can be very complex mixtures, involving a diverse plethora of chemical structures in a wide range of polarities, so that the restricted chromatographic resolution commonly limits the identification via MS to the more abundant compounds. Complex mixtures can be better resolved by employing comprehensive two-dimensional gas chromatography–mass spectrometry (GCxGC–MS), which has been defined as "...an orthogonal two-column separation, with complete transfer of a solute from the separation system 1 (column 1) to the separation system 2 (column 2), such that the separation performance from each system (column) is preserved" [64]. In GC × GC, two columns with different polarity—usually a nonpolar column in the first dimension and a moderately polar column for the second one—are run in series. Analytes eluting from the first dimension (1D) column are trapped, focused, and then rapidly injected, as a narrow band of few milliseconds, in the second dimension (2D) column, then the eluting peaks are detected by MS. The transfer process is actuated by a modulator, a thermal or valve-based focusing system. Each single modulator cycle takes a fixed time (4–8 s) and each fraction, injected online into the second column must be analyzed in a time equal to that of the successive modulation. The challenge is to avoid continuously transmitting analyte onto the second column, which would lead to a loss of resolution. A solution to this problem is to make the separation on the second column much faster than the separation on the first column. The volume of data generated is significantly larger than the one obtained in a one-dimensional analysis. However, this approach allows for better separation of the number of components in the sample. Although single qMS instruments are cheaper, can provide very low LODs via selected ion monitoring (SIM), and can provide maximum acquisition rates (20,000 amu/s) suitable for metabolic profiling, TOF has become the preferentially MS analyzers for GCxGC volatilome analysis. TOF-MS instruments are capable of full-spectrum collection rates up to 500 Hz with improved sensitivity. Besides the high-resolution mass spectrometry (HRMS) provide accurate mass data, which increases the identification confidence and allows to annotate molecular formulas for unknown compounds, being especially useful in untargeted metabolomic studies.

Metabolite identification remains a major complication. Although EI generates highly reproducible fragmentation spectra, only a relatively small percentage of metabolites can be identified by searching databases, mainly because these have traditionally been a repository of EI spectra of synthetic organic compounds. Only recently, the number of metabolite spectra started to increase. A more powerful identification method involves comparing both EI/CI spectra and retention indices obtained from analyzing a reference compound under identical analytical conditions. If commercial standards are not available, metabolite identification can be cumbersome.

Retention indexes (RI) were first introduced by Kováts [28] for isothermal analysis and then by Van den Dool [65] for temperature-programmed analysis (linear retention indices, LRIs) and are calculated vs. a homologous series of linear hydrocarbons run in the same GC conditions as samples. RI can also be automatically calculated using the Automated Mass Spectral Deconvolution and Identification System (AMDIS), freely available from the National Institute of Standards and Technology (NIST) at this site (http://www.amdis.net/).

In order to achieve the identification of unknown compounds, their background-subtracted EI spectra are searched against EI libraries (such as the NIST library) to achieve identification. Values of *m/z* values and relative ion intensities in a spectrum are matched against spectra in a reference library [the most used database of EI spectra is the NIST database (http://www.nist.gov/srd/nist1.htm)]. The database search will usually return a list of possible hits, ranked by the probability of the match. Although, even if a match is exceedingly high, the metabolite should still not be considered as identified.

The high variability of data obtained from the investigated matrix composition makes it hard to indicate a universal approach to quantitatively evaluate the volatilome composition. The most widely used approaches are: (a) relative percentage abundance, (b) internal standard normalized percentage abundance, and (c) "absolute" or true quantitation of one or more target components, with or without a validated method.

Relative percentage abundance can be applied only to evaluate relative component ratios within the same sample. Internal standard normalized percentage abundance is the ideal approach when a group of samples is compared: raw data must first be corrected vs. analyte response factors to the detector, then normalized vs. an internal standard. Percentage abundance must be calculated vs. the sum of the areas of a fixed number of selected components, found in all the samples. The quantitation of marker components is obtained from the chromatographic area in SIM mode vs. an internal (or external) standard and calculated via a calibration curve constructed from amounts of pure standards in the selected concentration range.

Some common non-separative techniques used in the study of volatilome using mass spectrometry are selected-ion flow-tube mass spectrometry (SIFT-MS) and proton-transfer-reaction mass spectrometry (PTR-MS). These techniques are focused on the use of soft chemical ionization, allow on-line detection of VOCs with low levels of detection without the need for pre-concentration or sample preparation, which facilitates obtaining reproducible results. For example, Vendel and co-workers [66], used SIFT-MS and HS-SPME-GC–MS for the analysis of strawberry aroma. Although both techniques provided similar results in the study of the fruit ripening, the SIFT-MS analysis was about 11 times faster than HS-SPME-GC–MS. Moreover, SIFT-MS showed low detection limits, so that the postharvest analysis can be easily performed by the analysis of individual fruit. Capellin and collaborators [67] developed a similar study was using PTR-TOF-MS to study the volatilome of clones belonging to three types of apple. They concluded that PTR-TOF-MS is a very useful tool for volatilome studies once this technique allows obtaining a rapid and non-invasive fingerprint of the VOCs profile from single apple fruits.

With an alternative focus, the chromatographic system can be coupled to an olfactometer detector to identify the aroma-active compounds present in a determinate volatilome. This type of analysis allows determining the compounds which generate a positive response to the electronic noise detector, obtaining their identification by comparison of the mass spectrum, retention index, and odor descriptions with reference compounds. Using gas chromatography-olfactometry-mass spectrometry (GC-O-MS), Zhu and co-workers [68] studied the volatile profile of three cultivars of mulberries, establishing benzaldehyde, ethyl butanoate, (E)-2-nonenal, 1-hexanol, hexanal, methional, 3-mercaptohexyl acetate, and 3-mercapto-1-hexanol as the main compounds responsible for the characteristic aroma of mulberry.

5. Volatilome data processing

Once the raw data have been acquired following chromatographic separation and mass spectrometry analysis, the large amount of data generated needs to be processed following a standardized procedure that includes data conversion, pre-processing, pre-treatment, and metabolite annotation [69]. An additional step, sharing data derived from any metabolomics analysis, currently is optional for researchers but highly recommended.

5.1 Extract raw files from instruments and proceed to data conversion

Data processing starts with a set of raw data files for different samples. Usually, default vendor formats from instruments need a conversion. A useful toolkit compatible with several instruments formats is ProteoWizard (http://proteowizard.sourceforge.net/download.html) [70]. Open-source formats usually supported by many software packages are Network Common Data Form (NetCDF) [71], Extensible Markup Language (mzXML) [72], and Mass Spectrometry Markup

Language (mzmL) [73]. Each file is processed to an easily accessible and more informative data table, where rows represent samples and columns represent different features from volatilome. Values from this matrix represent intensity values of peak area/height, standing for relative concentration. The data should be checked for missing values and possible outliers.

5.2 Set parameters to perform data pre-processing

Pre-processing involves setting different filters to recognize signals from noise, select masses or intensities to perform feature detection, and finally adjust the retention time shifts parameters needed to align features throughout all samples. The aim of pre-processing is to minimize the number of false positives features and to establish quantitative procedures for discarding less reliable signals with low signal-to-noise ratio, or low prevalence within a similar set of samples [74].

5.3 Choose the best method to perform data pre-treatment

Pre-treatment or data correction is one of the most important steps from data analysis because systematic and technical variation could obscure relevant biological patterns. The variation in the data resulting from a metabolomics experiment is the sum of the induced variation and the total uninduced variation [75]. Some sources of variation could be controlled by researchers through a careful experimental design. In other cases, this variation is very difficult to control. Natural variation in the metabolism of an organism can cause 5000-fold differences in signal intensities for different metabolites, or sampling could not be performed on the exact conditions for all samples, sample work-up varies naturally between batches, and analytical errors are always present. This variation could be accounted for using different classes of corrections that include centering, scaling, transformation, and normalization of raw data and several methods are available to do so (e.g., autoscaling, pareto scaling, range scaling, vast scaling, log transformation, and power transformation, normalization by sum, normalization by a reference sample). The selection of the most appropriate method depends on the hypothesis to be tested and the statistical behavior of the data matrix. Before applying pre-treatment methods, it is required to check if data is fit for analysis. For example, performing the treatment may enhance the results of a clustering method (if the hypothesis is related to comparison of similarities), while obscuring the results of a Principal Component Analysis (PCA) (if in contrast, the hypothesis is related with determining redundancy between metabolites) [75].

5.4 Metabolite annotation

The analysis by comparison with pure standards of different family of compounds is advisable, in order to compare the retention rates of the compounds. However, the characterization of a certain metabolite that there are no pure standards, its determination can be done by comparison with homologues of a certain family of compounds, which the detailed analysis of the fragmentation pattern. Metabolite annotation is still challenging despite all efforts made for establishing specialized databases with mass spectral properties of different metabolites. Annotation and identification levels for metabolites were defined by the Chemical Analysis Working Group of the Metabolomics Standards Initiative (MSI). Level 1 indicates compromise identified compounds, level 2 is used for putatively annotated compounds, level 3 is used for putatively characterized compound classes, and level 4 is used for unidentified or unclassified metabolites that still can be differentiated and quantified based upon spectral data. Dark matter, also called "unknown unknowns", represents the majority of metabolites analyzed on a metabolomics experiment, because instruments collect much more information than it is currently possible to annotate [76]. It is estimated that an average of only 2% of the data can be annotated. This is even a most common problem in metabolomics analysis from animals because many databases are specialized in human-derived metabolites, or some molecular structures from animals have been solved but are absent from the reference databases. Analysis from non-model organisms tends to have a higher number of truly novel compounds, called "unknown unknowns" [77]. As it is impossible to collect spectra for every molecule in the universe, computer-generated (in silico) spectral prediction algorithms are also recommended during metabolite annotation such as CSI:FingerID (https://www.csi-fingerid.uni-jena.de/) and Competitive Fragmentation Modeling-ID (CFM-ID, https://cfmid.wishartlab.com/) for analyzing fragmentation patterns. For volatilome analysis NIST (https://www.mswil.com/software/spectral-libraries-and-databases/nist20/) and Wiley (https://www.mswil.com/software/spectral-libraries-and-databases/wiley-spectral-libraries/wiley-gcms-libraries/) electronic collections are the most used mass spectra databases. The Dictionary of Natural Products (DNP) (http://dnp.chemnetbase.com/faces/chemical/ChemicalSearch.xhtml;jsessionid=DBE98AD72918A1607A7E739064D0DB21), Pherobase (https://www.pherobase.com/), Human Metabolome Database (HMDB) (https://hmdb.ca/), METLIN (https://metlin.scripps.edu/landing_page.php?pgcontent=mainPage), MassBank Japan (http://www.massbank.jp/), MassBank Europe (https://massbank.eu/MassBank/), MassBank North America (https://mona.fiehnlab.ucdavis.edu/), Supernatural II (http://bioinf-applied.charite.de/supernatural_new/index.php), ChEMBL (https://www.ebi.ac.uk/chembl/), Mass Spectral and GC Data of Drugs, Poisons, Pesticides, Pollutants, and Their Metabolites (https://www.wiley.com/en-gb/Mass+Spectral+and+GC+Data+of+Drugs%2C+Poisons%2C+Pesticides%2C+Pollutants%2C+and+Their+Metabolites%2C+5th+Edition-p-9783527342877) and vocBinBase (https://bitbucket.org/fiehnlab/binbase/src/master/) are other useful resources. When compound annotation is not possible and only chemical class could be assigned to a metabolite it is recommended to employ the comprehensive, and computable chemical taxonomy from Classyfire (http://classyfire.wishartlab.com/). See [78] for a review focused on mass spectral databases for LC/MS- and GC/MS-based metabolomics. For the analysis of mVOCs, in 2014 was developed a software that allows the characterization of mass spectra obtained in microorganisms. It was updated in 2018 with more than 2000 compounds from more than 1000 species, which is called mVOC database 2.0 (http://bioinformatics.charite.de/mvoc) [79]. With this tool a more precise characterization of the different volatilome of the microbes studied at present is achieved.

6. Select the best statistical analysis for the research question and coherent with data pre-treatment

Select the univariate statistics according to the variables of interest. T-test, U-test, and analysis of variance (ANOVA) are the most common univariate statistics employed for data mining in volatilomics. As datasets usually include a large number of features, the significance level should be determined appropriately to reduce the number of false positives and false negatives. For reducing false positive, family wise error rate (FWER) correction, such as a Bonferroni correction, is a conservative approach, in which the p-values are multiplied by the number of comparisons. In contrast, for reducing false negatives, false discovery rate (FDR) correction is a highly sensitive method [80].

7. Select the best suitable multivariate statistics

Multivariate statistical methods are very powerful at summarizing large and multidimensional data generated from volatilomics. Exactly as for pre-treatment methods, multivariable analysis should be chosen carefully and selected coherently with the hypothesis of interest and methods used for data pretreatment. Unsupervised approaches and supervised approaches differ in how samples are grouped within the multivariate calculations. Unsupervised solely have access to the matrix to find features useful for grouping and categorizing the samples. Clustering methods, such as hierarchical clustering (HCA), K-means clustering, self-organizing maps, principal component analysis (PCA) are among this group. Once the data have been analyzed by unsupervised methods, supervised methods (e.g. partial least squares discriminant analysis (PLD-DA), artificial neural networks, and evolutionary algorithms) should be applied for further evaluation [81]. Supervised methods have access to qualitative or quantitative traits (e.g., specie, location, body size, tissue type) and the matrix of measurements and can classify samples. Volcano plots have also recently been used to identify significantly covarying metabolites in binary comparisons. Volcano plots show each features' statistical significance, p-value, on the y-axis, and fold change along the x-axis [82].

7.1 Determine if network inference provide better insights about data interpretation

Correlation networks is a visualization tool that summarizes positive and negative correlations found between samples that represent different biological process [69]. Molecular networking organizes metabolite features from a volatilomics analysis into a connectivity network based on similarities in molecular fragmentation patterns obtained from mass spectrometry [82]. This analysis cluster families of molecules through vector correlations between fragment ions and enhance the interpretation of volatilome differentiation using a chemically informed visualization. Also, it enhances the annotation process with experimental and in silico databases [83]. When it is possible to combine Volatilomic and Genomic analysis, molecular networking can also be useful to prioritize features by linking observed natural products to their cognate biosynthetic gene clusters and gene cluster families [82].

7.2 Whenever possible, share data in public repositories

Recently, many researchers have shared raw data files on open repositories, and this has motivated computer scientists to develop modern algorithms for facilitating the comparison of MS spectra obtained in different conditions [78]. This comparison still needs human inspection from experts trained in mass spectrometry fragmentation patterns, because is not an automatic process. Some examples of sites that allow raw experimental data to be shared in public repositories include MetaboLights (http://www.ebi.ac.uk/metabolights/), the Metabolomics Workbench (https://www.metabolomicsworkbench.org/), XCMS Online (https://xcmsonline.scripps.edu/landing_page.php?pgcontent=mainPage), MetabolomeExpress (https://www.metabolome-express.org/), GNPS (https://gnps.ucsd.edu/ProteoSAFe/static/gnps-splash.jsp) and the Metabolomic Repository Bordeaux (http://services.cbib.u-bordeaux.fr/MERYB/).

8. Conclusions

Current technological advances in sample collection, extraction techniques, volatile profiling, and data processing allow that the analysis of an invisible world where VOCs mediates different ecological processes could recover a more accurate picture of the complex chemical communication that occurs in nature. Different combinations of procedures need to be followed by researchers with the aim to answer specific scientific questions or hypotheses. Microextraction techniques emerge as tools for increasing extraction efficiency and at the same time facilitating faster extraction times without the environmental impact of large volume solvent wastes. Gas chromatography has played a fundamental role to detect volatile compounds often present as trace levels. Mass spectrometry has proved to be the preferred technique for the structure elucidation of new compounds and annotation of known VOCs. Current improvements in data analysis allow to extract of more biologically relevant information from a single study and to standardize procedures for evaluating hypothesis properly. All these steps are of paramount importance to evaluate both the ecological function of these compounds and the economic value in the medical, agricultural, flavor, and fragrance industry.

Acknowledgements

The authors thank the Department of Chemistry and Vicerrectoria de Investigaciones at Universidad de los Andes, Bogotá, Colombia for financial support. We wish to thank to Ministerio de Ciencia, Tecnología e Innovación (MinCiencias) for Julie Paulin Garcia Rodriguez (No 679), Mabel Gonzalez (No 757) and Gerson-Dirceu López (No 785), as well as the support to No. 44842-058-2018 and No. 80740-532-2019 projects. Also, the Faculty of Sciences of the Universidad de los Andes forgivable loan and research funds (INV-2018-2033-1259, INV-2019-2067-1747, INV-2018-2048-1338, and INV-2019-2086-1843). Scholarship granted by Fulbright to Mabel González as a Visiting Scholar at the Dorrestein Laboratory at Skaggs School of Pharmacy & Pharmaceutical Sciences, University of California, San Diego, United States.

Conflict of interest

The authors declare no conflict of interest.

Abbreviations

1D	First-Dimension
2D	Second-Dimension
AC	Alternating Current
AMDIS	Automated Mass Spectral Deconvolution and Identification System
ANOVA	Analysis of variance
ATCC	American Type Culture Collection
CAR	Carboxen
CFM-ID	Competitive Fragmentation Modeling-Id
CI	Chemical Ionization
CITES	Convention on International Trade in Endangered Species of Wild Fauna and Flora

CW	Carbowax
DC	Direct Current
SDE	Simultaneous Extraction-Distillation
DHS	Dynamic Headspace
DNP	Dictionary of Natural Products
DVB	Divinylbenzene
EI	Electron Ionization
EOs	Essential Oils
FWER	Family Wise Error Rate
FDR	False Discovery Rate
GC	Gas Chromatography
GC–MS	Gas Chromatography–Mass Spectrometry
GC-O-MS	Gas Chromatography-Olfactometry-Mass Spectrometry
GNPS	Global Natural Products Social Networking
GCxGC	Comprehensive Two-Dimensional Gas Chromatography
HCA	Hierarchical Clustering
HF-LPME	Hollow Fiber Liquid-Phase Microextraction
HMDB	Human Metabolome Database
HRMS	High-Resolution Mass Spectrometry
HS	Headspace
IT-MS	Ion-Trap Mass Spectrometer
LLE	Liquid–Liquid Extraction
LPME	Liquid-Phase Microextractions
LRI	Linear Retention Indices
MS	Mass Spectrometer
MSI	Metabolomics Standards Initiative
mVOCs	Microbial Volatile Organic Compounds
m/z	Mass-To-Charge Ratio
mzmL	Mass Spectrometry Markup Language
mzXML	Extensible Markup Language
NetCDF	Network Common Data Form
NIST	National Institute of Standards and Technology
PA	Polyacrylate
PCA	Principal Component Analysis
PDMS	Polydimethylsiloxane
PLD-DA	Partial Least Squares Discriminant Analysis
PTR-MS	Proton-Transfer-Reaction Mass Spectrometry
PTV	Programmed Temperature Vaporizer
RI	Retention indexes
qMS	Quadrupole Mass Spectrometer
SAFE	Solvent Assisted Flavor Evaporation
SBSE	Stir Bar Sorptive Extraction
SDME	Single Drop Microextraction
SIFT-MS	Selected-Ion Flow-Tube Mass Spectrometry
SIM	Selected Ion Monitoring
SLM	Supported Liquid Membrane
SPE	Solid-Phase Extraction
SPME	Solid-Phase Microextraction
TOF-MS	Time-of-Flight Mass Spectrometer
VOCs	Volatile Organic Compounds

Author details

Chiara Carazzone*, Julie P.G. Rodríguez, Mabel Gonzalez and Gerson-Dirceu López
Laboratory of Advanced Analytical Techniques (LATNAP), University of los Andes, Bogotá D.C., Colombia

*Address all correspondence to: c.carazzone@uniandes.edu.co

References

[1] Crandall SG, Gold KM, Jiménez-Gasco M del M, Filgueiras CC, Willett DS. A multi-omics approach to solving problems in plant disease ecology. PLoS One. 2020 Sep 22;15(9):e0237975. DOI: 10.1371/journal.pone.0237975

[2] Raguso RA. Start making scents: the challenge of integrating chemistry into pollination ecology. Entomol Exp Appl. 2008 Jul 19;128(1):196-207. DOI: 10.1111/j.1570-7458.2008.00683.x

[3] Raguso ROEA, Pichersky E. Systematics and Evolution (Onagraceae): recent evolution of floral scent and moth pollination. America (NY). 1995;194:55-67. DOI: 10.1007/BF00983216

[4] Raguso RA. Floral scent in a whole-plant context: Moving beyond pollinator attraction. Funct Ecol. 2009 Oct;23(5):837-40. DOI:10.1111/j.1365-2435.2009.01643.x

[5] Raguso RA. Wake up and smell the roses: The ecology and evolution of floral scent. Annu Rev Ecol Evol Syst. 2008;39:549-69. DOI: 10.1146/annurev.ecolsys.38.091206.095601

[6] Fangel JU, Pedersen HL, Melgosa SV-, Ahl LI, Salmean AA, Egelund J, et al. High-Throughput Phenotyping in Plants. Normanly J, editor. Totowa, NJ: Humana Press; 2012. 351-362 p. (Methods in Molecular Biology; vol. 918). DOI: 10.1007/978-1-61779-995-2

[7] Learning M, Cookbook R. The Chemistry and Biology of Volatiles. Herrmann A, editor. Chichester, UK: John Wiley & Sons, Ltd; 2010. DOI: 10.1002/9780470669532

[8] Ceccarelli A, Spinelli F, Farneti B, Khomenko I, Cellini A, Donati I, et al. Postharvest Biology and Technology Nectarine volatilome response to fresh-cutting and storage. Postharvest Biol Technol. 2020;159(August 2019):111020. DOI: 10.1016/j.postharvbio.2019.111020

[9] Lytou AE, Panagou EZ, Nychas G-JE. Volatilomics for food quality and authentication. Curr Opin Food Sci. 2019 Aug;28:88-95. DOI: 10.1016/j.cofs.2019.10.003

[10] Rice S, Maurer D, Fennell A, Dharmadhikari M, Koziel J. Evaluation of Volatile Metabolites Emitted In-Vivo from Cold-Hardy Grapes during Ripening Using SPME and GC-MS: A Proof-of-Concept. Molecules. 2019 Feb;24(3):536. DOI: 10.3390/molecules24030536

[11] Chen X, Köllner TG, Jia Q, Norris A, Santhanam B, Rabe P, et al. Terpene synthase genes in eukaryotes beyond plants and fungi: Occurrence in social amoebae. Proc Natl Acad Sci. 2016 Oct;113(43):12132-7. DOI: 10.1073/pnas.1610379113

[12] Schulz-Bohm K, Martín-Sánchez L, Garbeva P. Microbial Volatiles: Small Molecules with an Important Role in Intra- and Inter-Kingdom Interactions. Front Microbiol. 2017 Dec;8(DEC):1-10. DOI: 10.3389/fmicb.2017.02484

[13] Gonçalves Barbosa JM, Machado Botelho AF, Santana da Silva RH, Ferreira de Almeida SS, Ferreira ER, Caetano David L, et al. Identification of cattle poisoning by Bifenthrin via earwax analysis by HS/GC–MS. Biomed Chromatogr. 2020 Nov;(August). DOI: 10.1002/bmc.5017

[14] Barbosa JMG, Fernandes Rodrigues MK, David LC, e Silva TC, Fortuna Lima DA, Pereira NZ, et al. A volatolomic approach using cerumen as biofluid to diagnose bovine intoxication

by *Stryphnodendron rotundifolium*. Biomed Chromatogr. 2020;34(11):1-11. DOI: 10.1002/bmc.4935

[15] Sardans J, Peñuelas J, Rivas-Ubach A. Ecological metabolomics: overview of current developments and future challenges. Chemoecology. 2011 Dec;21(4):191-225. DOI: 10.1007/s00049-011-0083-5

[16] Jarmusch AK, Wang M, Aceves CM, Advani RS, Aguirre S, Aksenov AA, et al. ReDU: a framework to find and reanalyze public mass spectrometry data. Nat Methods. 2020;17(9):901-4. DOI: 10.1038/s41592-020-0916-7

[17] Wang M, Carver JJ, Phelan V V., Sanchez LM, Garg N, Peng Y, et al. Sharing and community curation of mass spectrometry data with Global Natural Products Social Molecular Networking. Nat Biotechnol. 2016;34(8):828-37. DOI: 10.1038/nbt.3597

[18] Qualley A V., Dudareva N. Metabolomics of plant volatiles. In: Dmitry A. Belostotsky, editor. Methods in molecular biology (Clifton, NJ). Springer Science and Business Media LLC; 2009. p. 329-43. DOI: 10.1007/978-1-60327-563-7_17

[19] Sumner LW, Mendes P, Dixon RA. Plant metabolomics: Large-scale phytochemistry in the functional genomics era. Phytochemistry. 2003;62(6):817-36. DOI: 10.1016/S0031-9422(02)00708-2

[20] Iijima Y. Recent Advances in the Application of Metabolomics to Studies of Biogenic Volatile Organic Compounds (BVOC) Produced by Plant. Metabolites. 2014 Aug 21;4(3):699-721. DOI: 10.3390/metabo4030699

[21] Dudareva N, Negre F, Nagegowda DA, Orlova I. Plant volatiles: Recent advances and future perspectives. CRC Crit Rev Plant Sci. 2006 Oct 1;25(5):417-40. DOI: 10.1080/07352680600899973

[22] Bicchi C, Maffei M. The Plant Volatilome: Methods of Analysis. In: Normanly J, editor. High-Throughput Phenotyping in Plants: Methods and Protocols, Methods in Molecular Biology. Springer Science+Business Media; 2012. p. 289-310. DOI: 10.1007/978-1-61779-995-2

[23] Silva GL, Lee I-S, Kinghorn D. Special Problems with the Extraction of Plants. In: Cannel RJP, editor. Natural Products Isolation. Totowa: Humana Press Inc.; 1998. p. 343-63. DOI: 10.1007/978-1-59259-256-2_12

[24] Lisec J, Schauer N, Kopka J, Willmitzer L, Fernie AR. Gas chromatography mass spectrometry-based metabolite profiling in plants. Nat Protoc. 2006 Jun;1(1):387-96. DOI: 10.1038/nprot.2006.59

[25] Shuman JL, Cortes DF, Armenta JM, Pokrzywa RM, Mendes P, Shulaev V. Plant metabolomics by GC-MS and differential analysis. In: Pereira A, editor. Plant Reverse Genetics: Methods and Protocols, Methods in Molecular Biology. Springer Science+Business Media; 2011. p. 229-46. DOI: 10.1007/978-1-60761-682-5_17

[26] Knudsen JT, Eriksson R, Gershenzon J, St B. Diversity and Distribution of Floral Scent. Bot Rev. 2006;72(1):1-120. DOI: 10.1663/0006-8101(2006)72[1:DADOFS]2.0.CO;2

[27] Stashenko EE, Martínez JR. Sampling flower scent for chromatographic analysis. J Sep Sci. 2008 Jun;31(11):2022-31. DOI: 10.1002/jssc.200800151

[28] Vallone S, Lloyd NW, Ebeler SE, Zakharov F. Fruit volatile analysis using

an electronic nose. J Vis Exp. 2012;(61):1-7. DOI: 10.3791/3821

[29] Learning M, Cookbook R. The Chemistry and Biology of Volatiles. Herrmann A, editor. Chichester, UK, UK: John Wiley & Sons, Ltd; 2010. DOI: 10.1002/9780470669532

[30] Elmassry MM, Piechulla B. Volatilomes of Bacterial Infections in Humans. Front Neurosci. 2020 Mar 25;14:257. DOI: 10.3389/fnins.2020.00257

[31] Weisskopf L, Schulz S, Garbeva P. Microbial volatile organic compounds in intra-kingdom and inter-kingdom interactions. Nat Rev Microbiol. 2021 Feb 1; DOI: 10.1038/s41579-020-00508-1

[32] López G-D, Suesca E, Álvarez-Rivera G, Rosato A E, Ibáñez E, Cifuentes A, et al. Carotenogenesis of *Staphylococcus aureus*: New insights and impact on membrane biophysical properties. Biochim. Biophys. Acta - Mol. Cell Biol. Lipids, 2021 Aug 1866(8), 158941. DOI: 10.1016/j.bbalip.2021.158941

[33] Gonzalez M, Celis AM, Guevara-Suarez MI, Molina J, Carazzone C. Yeast smell like what they eat: Analysis of Volatile Organic Compounds of *Malassezia furfur* in Growth Media Supplemented with Different Lipids. Molecules. 2019 Jan 24;24(3):419. DOI: 10.3390/molecules24030419

[34] Tabares M, Ortiz M, Gonzalez M, Carazzone C, Vives Florez MJ, Molina J. Behavioral responses of *Rhodnius prolixus* to volatile organic compounds released in vitro by bacteria isolated from human facial skin. PLoS Negl Trop Dis. 2018 Apr 23;12(4):e0006423. DOI: 10.1371/journal.pntd.0006423

[35] Ferreira-Caliman MJ, Ribeiro Andrade-Silva AC, Guidetti-Campos MC, Casanova Turatti IC, Santos Do Nascimento F, Lopes NP. A non-lethal SPME method for insect cuticular analysis by GC-MS. Anal Methods. 2014;6(21):8823-8. DOI: 10.1039/c4ay01909a

[36] Barbosa-Cornelio R, Cantor F, Coy-Barrera E, Rodríguez D. Tools in the investigation of volatile semiochemicals on insects: From sampling to statistical analysis. Insects. 2019;10(8):1-35. DOI: 10.3390/insects10080241

[37] Smith BPC, Hayasaka Y, Tyler MJ, Williams BD. β-caryophyllene in the skin secretion of the Australian green tree frog, *Litoria caerulea*: An investigation of dietary sources. Aust J Zool. 2004;52(5):521-30. DOI: 10.1071/ZO04019

[38] Brunetti AE, Merib J, Carasek E, Caramão EB, Barbará J, Zini CA, et al. Frog volatile compounds: Application of in vivo SPME for the characterization of the odorous secretions from two species of Hypsiboas Treefrogs. J Chem Ecol. 2015;41(4):360-72. DOI: 10.1007/s10886-015-0564-z

[39] Rodriguez SA, Paliza ML, Nazareno MA. Influence of Adsorbent Nature on the Dynamic Headspace Study of Insect Semiochemicals. Aust J Chem. 2017;70(8):902-7. DOI: 10.1071/CH17064

[40] Tuckey NPL, Day JR, Miller MR. Determination of volatile compounds in New Zealand Greenshell™ mussels (*Perna canaliculus*) during chilled storage using solid phase microextraction gas chromatography-mass spectrometry. Food Chem. 2013;136(1):218-23. DOI: 10.1016/j.foodchem.2012.07.118

[41] Sato Y, Takagi S, Inomata E, Agatsuma Y. Odor-active volatile compounds from the gonads of the sea urchin Mesocentrotus nudus in the wild in Miyagi prefecture, Tohoku, Japan. Food Nutr Sci. 2019;10(07):860-75. DOI: 10.4236/fns.2019.107062

[42] Giordano G, Carbone M, Ciavatta ML, Silvano E, Gavagnin M, Garson MJ, et al. Volatile secondary metabolites as aposematic olfactory signals and defensive weapons in aquatic environments. Proc Natl Acad Sci U S A. 2017;114(13):3451-6. DOI: 10.1073/pnas.1614655114

[43] Birkemeyer CS, Thomsen R, Jänig S, Kücklich M, Slama A, Weiß BM, et al. Sampling the body odor of primates: Cotton swabs sample semivolatiles rather than volatiles. Chem Senses. 2016;41(6):525-35. DOI: 10.1093/chemse/bjw056

[44] Kean EF, Bruford MW, Russo IRM, Müller CT, Chadwick EA. Odour dialects among wild mammals. Sci Rep. 2017;7(1):1-6. DOI: 10.1038/s41598-017-12706-8

[45] Dehnhard M, Hildebrandt TB, Meerheim C, Valentine I, Göritz F. Chemical signals in giant panda urine (*Ailuropoda melanoleuca*). In: Chemical Signals in Vertebrates 13. Cham: Springer International Publishing; 2016. p. 363-79. DOI: 10.1007/978-3-319-22026-0_24

[46] Weiß BM, Marcillo A, Manser M, Holland R, Birkemeyer C, Widdig A. A non-invasive method for sampling the body odour of mammals. Methods Ecol Evol. 2017;9(2):420-9. DOI: 10.1111/2041-210X.12888

[47] Cumeras R, Cheung WHK, Gulland F, Goley D, Davis CE. Chemical analysis of whale breath volatiles: A case study for non-invasive field health diagnostics of marine mammals. Metabolites. 2014;4(3):790-806. DOI: 10.3390/metabo4030790

[48] Schivo M, Aksenov AA, Pasamontes A, Cumeras R, Weisker S, Oberbauer AM, et al. A rabbit model for assessment of volatile metabolite changes observed from skin: A pressure ulcer case study. J Breath Res. 2017;11(1). DOI: 10.1088/1752-7163/aa51d7

[49] Soini HA, Whittaker DJ, Wiesler D, Ketterson ED, Novotny M V. Chemosignaling diversity in songbirds: Chromatographic profiling of preen oil volatiles in different species. J Chromatogr A. 2013;1317:186-92. DOI: 10.1016/j.chroma.2013.08.006

[50] Shaw CL, Rutter JE, Austin AL, Garvin MC, Whelan RJ. Volatile and semivolatile compounds in gray catbird uropygial secretions vary with age and between breeding and wintering grounds. J Chem Ecol. 2011;37(4):329-39. DOI: 10.1007/s10886-011-9931-6

[51] García-Rubio S, Attygalle AB, Weldon PJ, Meinwald J. Reptilian chemistry: Volatile compounds from paracloacal glands of the american crocodile (*Crocodylus acutus*). J Chem Ecol. 2002;28(4):769-81. DOI: 10.1023/A:1015288726605

[52] Pruett JA, Zúñiga-Vega JJ, Campos SM, Soini HA, Novotny M V., Vital-García C, et al. Evolutionary interactions between visual and chemical signals: Chemosignals compensate for the loss of a visual signal in male *Sceloporus* lizards. J Chem Ecol. 2016;42(11):1164-74. DOI: 10.1007/s10886-016-0778-8

[53] Huang S, Xu J, Wu J, Hong H, Chen G, Jiang R, et al. Rapid detection of five anesthetics in tilapias by in vivo solid phase microextraction coupling with gas chromatography-mass spectrometry. Talanta. 2017;168(December 2016):263-8. DOI: 10.1016/j.talanta.2017.03.045

[54] Smith BP, Zini CA, Pawliszyn J, Tyler MJ, Hayasaka Y, Williams B, et al. Solid-phase microextraction as a tool for studying volatile compounds in frog skin. Chem Ecol. 2000;17(3):215-25. DOI: 10.1080/02757540008037674

[55] Smith BPC, Tyler MJ, Wllliams BD, Hayasaka Y. Chemical and olfactory characterization of odorous compounds and their precursors in the parotoid gland secretion of the green tree frog, *Litoria caerulea*. J Chem Ecol. 2003;29(9):2085-100. DOI: 10.1023/A:1025686418909

[56] Umile TP, McLaughlin PJ, Johnson KR, Honarvar S, Blackman AL, Burzynski EA, et al. Nonlethal amphibian skin swabbing of cutaneous natural products for HPLC fingerprinting. Anal Methods. 2014;6(10):3277-84. DOI: 10.1039/C4AY00566J

[57] Poth D, Wollenberg KC, Vences M, Schulz S. Volatile amphibian pheromones: Macrolides from mantellid frogs from madagascar. Angew Chemie - Int Ed. 2012;51(9):2187-90. DOI: 10.1002/anie.201106592

[58] Bojko B, Reyes-Garcés N, Bessonneau V, Goryński K, Mousavi F, Souza Silva EA, et al. Solid-phase microextraction in metabolomics. TrAC - Trends Anal Chem. 2014 Oct 1;61:168-80. DOI: 10.1016/j.trac.2014.07.005

[59] Soares da Silva Burato J, Vargas Medina DA, Toffoli AL, Vasconcelos Soares Maciel E, Mauro Lanças F. Recent advances and trends in miniaturized sample preparation techniques. J Sep Sci. 2020 Jan 23;43(1):202-25. DOI: 10.1002/jssc.201900776

[60] Rice S, Maurer D, Fennell A, Dharmadhikari M, Koziel J. Evaluation of Volatile Metabolites Emitted In-Vivo from Cold-Hardy Grapes during Ripening Using SPME and GC-MS: A Proof-of-Concept. Molecules. 2019 Feb 1;24(3):536. DOI:

[61] Feng S, Huang M, Crane JH, Wang Y. Characterization of key aroma-active compounds in lychee (*Litchi chinensis Sonn.*). J Food Drug Anal. 2018 Apr;26(2):497-503. DOI: 10.1016/j.jfda.2017.07.013

[62] Riazanskaia S, Blackburn G, Harker M, Taylor D, Thomas CLP. The analytical utility of thermally desorbed polydimethylsilicone membranes for in-vivo sampling of volatile organic compounds in and on human skin. Analyst. 2008;133(8):1020-7. DOI: 10.1039/b802515k

[63] Sangster T, Major H, Plumb R, Wilson AJ, Wilson ID. A pragmatic and readily implemented quality control strategy for HPLC-MS and GC-MS-based metabonomic analysis. Analyst. 2006;131(10):1075-8. DOI: 10.1039/b604498k

[64] Giddings JC. Concepts and Comparisons in Multidimensional Separation. J High Resolut Chromatogr. 1987;10(5):319-23. DOI: 10.1002/jhrc.1240100517

[65] Arov DZ, Dym H. A generalization of the retention index system including linear temperature programmed gas—liquid partition chromatography. Oper Theory Adv Appl. 1963;11:463-71. DOI: 10.1007/978-3-319-70262-9_7

[66] Vendel I, Hertog M, Nicolaï B. Fast analysis of strawberry aroma using SIFT-MS: A new technique in postharvest research. Postharvest Biol Technol. 2019 Jun;152(June 2019):127-38. DOI: 10.1016/j.postharvbio.2019.03.007

[67] Cappellin L, Soukoulis C, Aprea E, Granitto P, Dallabetta N, Costa F, et al. PTR-ToF-MS and data mining methods: a new tool for fruit metabolomics. Metabolomics. 2012 Oct 4;8(5):761-70. DOI: 10.1007/s11306-012-0405-9

[68] Zhu JC, Wang LY, Xiao ZB, Niu YW. Characterization of the key aroma compounds in mulberry fruits by application of gas

chromatography–olfactometry (GC-O), odor activity value (OAV), gas chromatography-mass spectrometry (GC–MS) and flame photometric detection (FPD). Food Chem. 2018;245(August 2017):775-85. DOI: 10.1016/j.foodchem.2017.11.112

[69] Aksenov AA, Da Silva R, Knight R, Lopes NP, Dorrestein PC. Global chemical analysis of biology by mass spectrometry. Nat Rev Chem. 2017;1(October). DOI: 10.1038/s41570-017-0054

[70] Orchard S, Hoogland C, Bairoch A, Eisenacher M, Kraus HJ, Binz PA. Managing the data explosion: A report on the HUPO-PSI workshop august 2008, amsterdam, the Netherlands. Proteomics. 2009;9(3):499-501. DOI: 10.1002/pmic.200800838

[71] Rew R, Davis G. NetCDF: an interface for scientific data access. IEEE Comput Graph Appl. 1990 Jul;10(4):76-82. DOI: 10.1038/nbt1031

[72] Pedrioli PGA, Eng JK, Hubley R, Vogelzang M, Deutsch EW, Raught B, et al. A common open representation of mass spectrometry data and its application to proteomics research. Nat Biotechnol. 2004;22(11):1459-66. DOI: 10.1038/nbt1031

[73] Chambers MC, MacLean B, Burke R, Amodei D, Ruderman DL, Neumann S, et al. A cross-platform toolkit for mass spectrometry and proteomics. Nat Biotechnol. 2012;30(10):918-20. DOI: 10.1038/nbt.2377

[74] Eliasson M, Rännar S, Trygg J. From data processing to multivariate validation--essential steps in extracting interpretable information from metabolomics data. Curr Pharm Biotechnol. 2011;12(7):996-1004. DOI: 10.2174/138920111795909041

[75] van den Berg RA, Hoefsloot HCJ, Westerhuis JA, Smilde AK, van der Werf MJ. Centering, scaling, and transformations: improving the biological information content of metabolomics data. BMC Genomics. 2006 Jun 8;7(19):142. DOI: 10.1186/1471-2164-7-142

[76] Da Silva RR, Dorrestein PC, Quinn RA. Illuminating the dark matter in metabolomics. Proc Natl Acad Sci U S A. 2015;112(41):12549-50. DOI: 10.1073/pnas.1516878112

[77] Peters K, Worrich A, Weinhold A, Alka O, Balcke G, Birkemeyer C, et al. Current challenges in plant Eco-Metabolomics. Int J Mol Sci. 2018;19(5):1-38. DOI: 10.3390/ijms19051385

[78] Vinaixa M, Schymanski EL, Neumann S, Navarro M, Salek RM, Yanes O. Mass spectral databases for LC/MS- and GC/MS-based metabolomics: State of the field and future prospects. TrAC - Trends Anal Chem. 2016;78:23-35. DOI: 10.1016/j.trac.2015.09.005

[79] Lemfack MC, Gohlke BO, Toguem SMT, Preissner S, Piechulla B, Preissner R. MVOC 2.0: A database of microbial volatiles. Nucleic Acids Res. 2018 Jan 1;46:D1261-5. DOI: 10.1093/nar/gkx1016

[80] Putri SP, Fukusaki E. Mass Spectrometry-Based Metabolomics. A Practical Guide. CRC Press. Boca Raton; 2014. 294 p. DOI: 10.1039/c6np00048g

[81] Villas-Bôas SG, Mas S, Åkesson M, Smedsgaard J, Nielsen J. Mass spectrometry in metabolome analysis. Vol. 24, Mass Spectrometry Reviews. 2005. p. 613-46. DOI: 10.1002/mas.20032

[82] Covington BC, McLean JA, Bachmann BO. Comparative mass spectrometry-based metabolomics strategies for the investigation of

microbial secondary metabolites. Nat Prod Rep. 2017;34(1):6-24. DOI: 10.1039/c6np00048g

[83] Watrous J, Roach P, Alexandrov T, Heath BS, Yang JY, Kersten RD, et al. Mass spectral molecular networking of living microbial colonies. Proc Natl Acad Sci U S A. 2012;109(26):1743-52. DOI: 10.1073/pnas.1203689109

Chapter 5

Seagrass Metabolomics: A New Insight towards Marine Based Drug Discovery

Danaraj Jeyapragash, Ayyappan Saravanakumar and Mariasingarayan Yosuva

Abstract

Metabolomics is one of the new field of "Omics" approach and the youngest triad of system biology, which provides a broad prospective of how metabolic networks are controlled and indeed emerged as a complementary tool to functional genomics with well-established technologies for genomics, transcriptomics and proteomics. Though, metabolite profiling has been carried out for decades, owing to decisive mechanism of a molecule regulation, the importance of some metabolites in human regimen and their use as diagnostic markers is now being recognized. Plant metabolomics therefore aims to highlight the characterization of metabolite pool of a plant tissue in response to its environment. Seagrassses, a paraphyletic group of marine hydrophilous angiosperms which evolved three to four times from land plants back to the sea. Seagrasses share a number of analogous acquired metabolic adaptations owing to their convergent evolution, but their secondary metabolism varied among the four families that can be considered as true seagrasses. From a chemotaxonomic point of view, numerous specialized metabolites have often been studied in seagrasses. Hence, this chapter focus the metabolome of seagrasses in order to explore their bioactive properties and the recent advancements adopted in analytical technology platforms to study the non-targeted metabolomics of seagrasses using OMICS approach.

Keywords: seagrass, metabolomics, OMICS, non-targeted, drug discovery

1. Introduction

Over the past decades, metabolomics has emerged as a valuable tool for the comprehensive profiling and metabolic networks in the biological system. Pauling et al. [1] coined the term metabolomics which was first used in 1998 and even up to 2010 metabolomics was considered as an emerging one in the science field. Reports were documented on the complete genome ([2]; Yu et al., 2002), transcriptome [3] and proteome studies [4–6], but in recent years metabolome analyses using mass spectrometry (MS) - based platforms attracted attention. Even though, metabolite profiling have been carried out for decades, due to ultimate mechanism of a molecule regulation as constituents of metabolic pathways, the prominence of some metabolites in human regimen and their use as diagnostic markers is now being recognized [7].

Currently, metabolomics is a powerful tool for characterizing the metabolites and their metabolic pathways which provides a clear metabolic picture of biological samples. Metabolites are small molecules with diverse structures that are chemically transformed during the cellular metabolism [8]. The number of metabolites is expected to be significantly lowered than the number of genes, mRNAs and proteins which reduce the sample complexity. So far, the total number of metabolites in the plant kingdom is estimated to exist between 100000 to 200000, which make the task more challenging to detect more diverse group of metabolites [9]. Plant metabolomics therefore aims to highlight the characterization of metabolite pool of a plant tissue in response to its environment [10–13]. Since, metabolomics is a balanced approach that obtains inclusive information on the cell's, tissues or organisms metabolite content with low molecular weight, their configuration likely to be changed owing to diverse environmental conditions which reproduces different genetic background [14, 15].

Recent reports on the plant metabolome bought huge challenges to analytical technologies that have been used in current plant metabolomics programs. Some analytical approaches comprise metabolite profiling, metabolite target analysis and metabolite fingerprinting which can be employed according to focus of the research and research questions [16, 17]. Metabolite profiling does not certainly determine the absolute concentrations of metabolites; rather their comparative levels within a structurally related predefined group. Targeted metabolite analysis aims to determine the absolute concentration of metabolites using specialized extraction protocols with an adapted separation and detection methods [18]. Metabolite fingerprinting generally not used to detect individual metabolites, but rather it provides a fingerprint of all compounds which can be measured for sample comparison and discrimination analysis by non-specific rapid analysis of crude metabolite mixtures. However, single analytical technology is not enough to cover the whole metabolome owing to the metabolic diversity and their broad dynamic range in cellular abundance. Accordingly, different extraction techniques and combinations of analytical methods are often employed in order to acquire diverse group of metabolite coverage.

2. Mass spectrometry-based metabolomics analysis

Historically, metabolite concentrations were achieved either by spectrophotometric assays capable of detecting single metabolites or by simple chromatographic separation of mixtures with low complexity. However, over the past decade several methods with high accuracy and sensitivity have been established for the analysis of highly complex mixtures of compounds [19–21]. These methods include gas chromatography - mass spectrometry (GC–MS), liquid chromatography - mass spectrometry (LC–MS), fourier transform ion cyclotron resonance mass spectrometry (FT-ICR-MS) and capillary electrophoresis - mass spectrometry (CE-MS). In addition, NMR coupled with chromatography have found great efficacy in addressing specific issues with respect to medical fields [22, 23] and conceivably more important to the unequivocal determination of metabolite structures [24]. However, NMR shows relatively low sensitivity and hence can be used for profiling the diverse group of metabolites from complex mixtures. The pros and cons of mass-spectrometric based metabolomics is given in **Table 1**.

Gas Chromatography - Mass Spectrometry assists the identification and robust quantification of few hundred metabolites in a single plant extracts, which results in inclusive coverage of the central pathways of primary metabolism [25]. GC–MS has a major advantage than other methods that it has long been used for profiling the

Metabolomics Technology	Advantages	Disadvantages
GC–MS	• Well established chromatographic-mass spectrometric technology • Less expensive • Easy to perform • Better stability and reproducibility • Universal database for metabolite identification is available	• Time consuming • Can only analyze volatile compounds upon derivatization • Not able to identify the unknown compounds
LC–MS	• Highly sensitive • Method for sample pre-treatment is simple • Wide coverage of metabolite could be identified • Scanning range for different ions is usually less • Different columns can be used depends on the polarity of compounds	• Less reproducibility • Highly Expensive • Difficult to identify highly polar and charged metabolites
FT-ICR-MS	• High resolution mass spectrometric technique • Extemely high mass accuracy • High acquisition rates • Highly flexible	• Involves series of steps • Use large super conducting magnets • Highly expensive than GC–MS and LC–MS

Table 1.
Advantages and disadvantages of mass-spectrometric based metabolomics.

metabolites and therefore it has stable protocols for machine setup, their maintenance with chromatogram evaluation and interpretation. Though, single analytical system cannot cover the whole metabolome, GC–MS has a quite broad coverage of compounds classes including organic and amino acids, sugars, sugar alcohols, lipophilic compounds and phosphorylated intermediates [26]. During method validation, recovery experiments of all measurable compounds have been done and for unknown compounds, recombination experiments were executed to determine the recovery rates in which the extracts of two plant species are evaluated independently and also with mixtures [27, 28]. Liquid chromatography-based methods offer numerous advantages such as detection of broad range of metabolites, as they suffer from the lower reproducibility of retention time. In addition, they are more susceptible to ion suppression effects due to the predominant use of electrospray ionization, which renders the precise quantification more difficult [29–31].

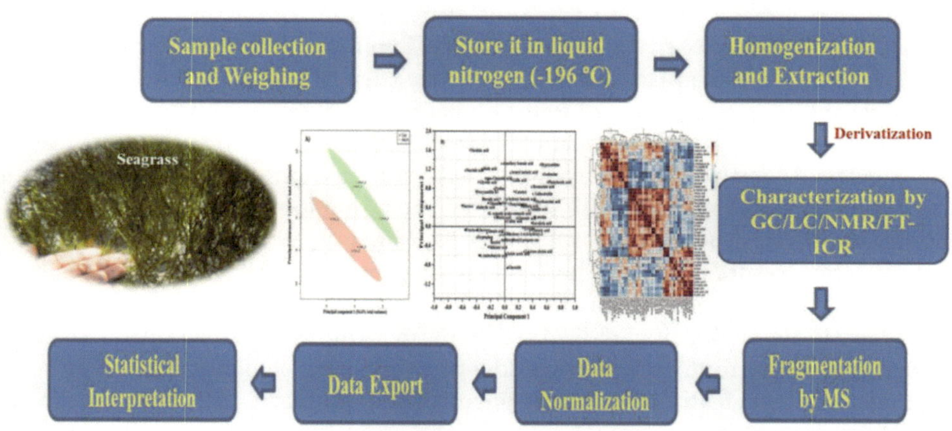

Figure 1.
Schematic overview of MS based metabolomics.

FT-ICR-MS and CE-MS has been reported to be worth mentioning, where FT-ICR-MS has unsurpassed mass accuracy thereby allows the researcher to obtain an idea about the chemical composition of the specific compounds. In case of CE-MS, the low-abundance metabolites can be detected and affords good chromatographic separation [32, 33]. Of these techniques, GC–MS is mostly preferred for the separation of low molecular weight metabolites which can be either volatile or can be converted into volatile and thermally stable compounds via chemical derivatization prior to the analysis [34]. The experimental procedure for GC–MS based metabolomics analysis is represented in **Figure 1**.

3. Derivatization

Derivatization is a process by which a compound is chemically modified to produce a new compound that has properties which are more amenable to specific analytical procedure. Samples analyzed by gas chromatography requires derivatization in order to make them suitable for analysis. Derivatization procedure imparts volatility, decreases the adsorption in the injector, increases the stability of compounds; improve the resolution and detectability between coeluting compounds and overlapping which assist in structure determination [35]. A good derivatizing reagents and the procedure should produce the compound of interest with desired chemical modification and be efficient, reproducible and non-hazardous (www.piercenet.com). For GC, derivatization reaction can be done by three basic types: silylation, acylation and alkylation. Silylating reagents react with compounds containing active hydrogen and are most frequently used in GC. Acylating reagents react with compounds having high polar functional groups such as amino acids or carbohydrates. While alkylating reagents target the active hydrogen's on amines and acidic hydroxyl group [36].

4. Seagrasses

Seagrasses, a marine hydrophilus angiosperm live entirely in an estuarine or in the marine environment and nowhere else [37]. Seagrass ecosystem act as a breeding and nursery ground for numerous organisms and also help in promoting the commercial fisheries. It is considered to be one of the most productive ecosystems that retain the structural complexity and biodiversity shed light to some researchers to describe seagrass community as marine representation of the tropical rainforests [38]. Currently, seagrasses are assigned to four families Hydrocharitaceae, Cymodoceaceae, Posidoniaceae and Zosteraceae (den [39, 40]). According to angiosperm Group III System, all four families occurred exclusively to monocot order Alismatales [41]; while Les and Tippery [40] favored to treat the same clades as a subclass Alismatidae. The family hydrocharitaceae comprises of three genera namely *Enhalus, Halophila and Thalassia* and the Cymodoceaceae encompasses the highest variety of genera includes Amphibolis, *Halodule, Syringodium, Cymodocea* and *Thalassodendron*. Likewise, the Zosteraceae comprises of *Zostera, Heterozostera, Nanozostera, Phyllospadix* which are exclusive marine organisms and the family Posidniaceae are monogeneric [42–44]. Therefore, seagrasses can be considered as a unique ecological group occurring worldwide in different climatic zones and sharing their metabolic features with their terrestrial counterparts [37]. However, their metabolism must have undergone several adaptations to survive and colonize in shorelines and oceans worldwide [37]. The seagrass ecosystems and their functions were given in **Table 2**.

Seagrass and Seagrass beds	Properties
Common names	Eelgrass, turtle grass, tape grass, shoal grass, and spoon grass
Families: 4	• Hydrocharitaceae • Posidonaceae • Zosteraceae • Cymodoceaceae
Total Species: 72	In India: 14 species exist
Habitat	Found in salt and brackish water
Depth	1 meter - 58-meter depth
Seagrass Parts	
Leaves • Chloroplast • Lacunae • Cuticle • Viens	Photosynthesis Help for buyoncy Exchange oxygen and carbondioxide in the water column Transport nutrients throughout the plants
Rhizome	Stabilize the seagrass beds under wave action
Roots	Absorb nutrients from the soil and transport to the plants
Growth and reproduction	Sexual reproduction and asexual clonal growth
Biodiversity • Small bodied	Small thin leaves Small rhizome "Guerilla" strategy Short lived with fast turnover Low biomass Abundant flowering Many small seeds and seed bank
• Large bodied	Large thick leaves Large rhizome "Phalanx" strategy Long-lived with slow turnover High biomass and holds space Patchy flowering Few large seeds and seeds germinate rapidly
Ecosystem benefits	Lungs of the sea Creation of Living Habitat Foundation of Coastal Food Webs Blue Carbon

Table 2.
Seagrass ecosystem and their benefits.

5. Seagrass metabolome

Seagrasses share a number of analogous acquired metabolic adaptations owing to their convergent evolution, but their secondary metabolism varies among the four families that can be considered as true seagrasses. During the period of ancient Tethys Sea, approximately 90 million years ago surrounded by Africa, Gondwana-land, and Asia, the terrestrial like species returned to the sea and thus explaining the "terrestrial-like" chemical profile of seagrass. From a chemotaxonomic viewpoint, numerous secondary metabolites have been often studied in seagrasses. The metabolome of seagrasses may differ with respect to geographical location, sub-strates and other physiological factors includes wide fluctuations in the salinity which are prone to synthesize novel metabolites with defined physiological, bio-chemical, defense and ecological roles [45, 46]. Preliminary suggestions confirmed that seagrasses have pharmaceutically potent bioactive secondary metabolites [47], that are directed to prove to be a lead molecule for drug discovery [48]. The status of metabolomic study in seagrasses reported so far is tabulated in **Table 3**.

Seagrass	Methods used	Derivation method	Results	Potential application	Reference
Zostera marina *Zostera noltii*	GC/TOF	Trimethyl silylation	Adaptivemechanisms are involved through metabolic pathways to dampen the impacts of heat stress	Sucrose, fructose, and myo-inositol were identified to be the most responsive metabolites of the 29 analyzed organic metabolites.	Gu et al. [49]
Cymodocea nodosa	GC-QTOF-MS	Trimethyl silylation	Growth promoting metabolites (sucrose, fructose, myo-inositol, heptacosane, tetracosane, stigmasterol, catechin and alpha-tocopherol) were lower close to the zone, whereas metabolites involved with stress-response (alanine, serine, proline, putrescine, ornithine, 3,4-dihydroxybenzoic acid and cinnamic acid) were identified	Metabolomic fingerprinting of seagrass provides opportunities for early detection of environmental degradation in marine ecological studies	Kock et al. [50]
Halodule pinifolia	GC–MS	Trimethylsilyl etherification	GC–MS analysis revealed the presence of thirty-five compounds which include flavonoids, sugars, amino acids and plant hormones	Study has explored a newer marine source, *H. pinifolia* for RA, which is an emerging potential preclinical chemical entity	Jeyapragash et al. [51]
Zostera marina	GC–MS	Trimethyl silylation	Decreased carbohydrate decomposition products and tricarboxylic acid (TCA) cycle intermediate products, indicating that the energy supply of the eelgrass may be insufficient at high temperature	composition of the membrane system of eelgrass may change at high temperature and implying that high temperature may cause the membrane system to be unstable	Gao et al. [52]
Halodule pinifolia	GC–MS	Trimethyl silylation	98 metabolites in wild and 125 metabolites in SCC were identified. 77 primary and secondary metabolism pathways in wild, while 73 metabolism pathways in SCC were reported	Baseline information on *H.pinifolia* metabo-lism in the marine and artificial environments	Jeyapragash et al. [53, 54]

Seagrass	Methods used	Derivation method	Results	Potential application	Reference
Zostera muelleri	NMR	Trimethyl-silylation	Several potential bioindicators of low-light stress: a reduction of soluble sugars and their derivatives, glucose, fructose, sucrose and myo-inositol, N-methylnicotinamide, organic acids and various phenolic compounds were identified	Metabolomics measurements may be useful bio-indicators of low-light stress in seagrass	Griffith et al. [55]
Halodule pinifolia	GC–MS	Trimethyl-Silylation	Three thermo-protective metabolites such as trehalose (sugar), glycine betaine (amino acid) and methyl vinyl ketone (organic acid) were profiled from *H. pinifolia* (45°C) and is the first report on the occurrence of glycine betaine and methyl vinyl ketone from seagrasses and other aquatic species so far	Facilitate the further research on identifying gene to metabolite networks for an effective management of seagrass conservation by genetic manipulation	Jeyapragash et al. (2021)

Table 3.
Status of metabolomics studies in seagrasses.

Primary metabolites from seagrasses reported to be similar, to that of any other terrestrial angiosperms [56]. Despite the higher phenol content, seagrasses found to be rich source of protein which alleviates the chronic problem of protein deficiency in developing countries like India [47]. In addition, seagrasses are a rich source of secondary metabolites such as simple phenolic compounds, phenylmethane and phenylethane derivatives, flavonoid and volatile derivatives with high commercial value [38]. Jeyapragash et al., reported that the plant growth regulators enhance the production of flavonoid production in the callus and cellular suspension cultured cells of seagrass *H. pinifolia* (**Figure 2**) Though, quantum of research has been published with respect to metabolome of seagrasses, there is still a dearth of knowledge as compared to the terrestrial plants. Seagrass offers remarkable opportunities to derive new commercially valuable phytochemicals when compared to algae [57, 58]. Therefore, the metabolite content from seagrasses constitutes another treasure of the ocean. Henceforth, the knowledge on metabolomics analysis from seagrass is decisive for understanding the complete metabolite picture and to explore their bioactive properties.

Seagrasses, the only higher plants solely living in the marine habitats and are ultimate importance for marine ecological systems close to the shorelines. Several studies dealt with the function of seagrasses as primary producers, shelter and food for fish, turtles and invertebrates as well as spawning areas for these organisms [59–61]. The reviews existing on seagrasses with different focus than the present one deal in more detail with other aspects of the ecological role of seagrasses,

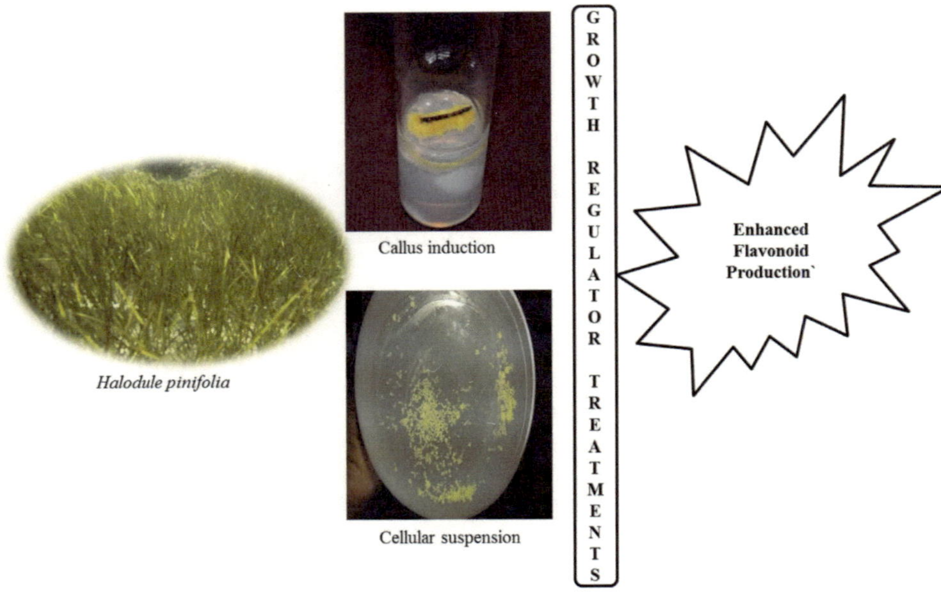

Figure 2.
Growth regulators mediated flavonoid production in callus and cellular siuspenison of H. pinifolia.

particulary the metabolite classes which are very few and primitive. Seagrasses reported to share the most features of primary and secondary metabolites with respect from the Alismatales order which live in land and freshwater habitats [62]. Kannan and Kannal [63] and Pradheeba et al. [56] reported that primary metabolites such as carbohydrate, protein and lipid content from seagrasses acts as a rich source of nutritional value and was eveidenced by the obvious increase in the carbohydrate content *of E. acoroides, T. hemporichii* [63], *C. rotundata* [56] and lipid content of *C. nodosa* [64]. Higher protein content form seagrasses have also been reported in *E. acoroides* [63], *Ruppia cirrhosa* [64], *C. serrulata, S. isoetifolium, H. ovalis, H. pinifolia* [47], *C. rotundata* and *H. uninervis* [56] that are found to accumulate manximum concentration than the seaweeds [47].

Secondary metabolism occurs in seagrasses depends on the season and environmental conditions and was reported as a rich source of diverse natural products from simple to conjugated phenolic compounds such as phenolic acids, flavones, tannins and lignins [65, 66]. It was also reported that *P. oceanica* harbors compounds ranging from simple phenol derivatives, phenylmethane, phenyethane and phenypropane derivatives [67]. Athiperumalsami *et al.* [47] reported the occurrence and absence of alkaloids, antroquinones, catechins, coumarins, flavonoids, phenols, saponins, quinones, tannins and secondary metabolites in the tropical seagrass *Cymodocea serrulata*, *Halodule pinifolia* and *Syringodium isoetifolium*. Heglmier and Zindron [62] reported 51 natural products in *P. oceanica* that includes phenols, flavones, phenylmethane, ethane derivatives. Though reports availed on the seagrass derivatives, limited study dealt with the chemistry of secondary metabolites of seagrasses [62, 68]. The list of metabolites profiled and the metabolic pathways from wild seagrasses are tabulated in **Table 4** and **Figure 3**.

The presence of sulphated flavones was reported to be accumulated in *Halophila* and *Thalassia* species [69] and in *Z. marina* [70], but not found in *Syringodium* spp. and *P. oceanica*. Mc Millan [71] found the presence of either flavones or phenolic acid in 43 segrass species while the sulphated flavones found specifically in five species namely, *Zostera, Phyllospadix, Enhallus, Thallasia* and *Halophila* species.

Compound Name	Molecular Formula	Molecular weight (g/ mol)	Exact Mass (g/mol)
D-Glucose	$C_6H_{12}O_6$	180.156	180.063
Maltose	$C_{12}H_{22}O_{11}$	342.297	342.116
D-Fructose	$C_6H_{12}O_6$	180.156	180.063
Sucrose	$C_{12}H_{22}O_{11}$	342.297	342.116
Inositol	$C_6H_{12}O_6$	180.156	180.063
Methyl alpha-D-Glucopyranose	$C_7H_{14}O_6$	194.183	194.079
D-Galactose	$C_6H_{12}O_6$	180.156	180.063
Lactose	$C_{12}H_{22}O_{11}$	342.297	342.116
L-Rhamnose	$C_6H_{12}O_5$	164.157	164.068
D-Ribose	$C_5H_{10}O_5$	150.13	150.053
Adenosine-2′:3′- cyclic monophosphate	$C_{10}H_{14}N_5O_7P$	347.224	347.063
N-Acetyl-Î-D-glucosamine	$C_8H_{15}NO_6$	221.209	221.09
Aspartyl-Leucine	$C_{10}H_{18}N_2O_5$	246.263	246.122
Glycine	$C_2H_5NO_2$	75.067	75.032
Threonine	$C_4H_9NO_3$	119.12	119.058
Valine	$C_5H_{11}NO_2$	117.148	117.079
Proline	$C_5H_9NO_2$	115.132	115.063
Alanine	$C_3H_7NO_2$	89.094	89.048
Thiamine	$C_{12}H_{17}N4OS^+$	265.355	265.112
Methionine	$C_5H_{11}NO_2S$	149.208	149.051
Phenylanaline	$C_9H_{11}NO_2$	165.192	165.079
Tyrosine	$C_9H_{11}NO_3$	181.191	181.074
Methyl Pyroglutamate	$C_6H_9NO_3$	143.142	143.058
Glutamic acid	$C_5H_9NO_4$	147.13	147.053
Vanillic acid	$C_8H_8O_4$	168.148	168.042
Oxalic acid	$C_2H_2O_4$	90.034	89.995
gamma-Aminobutyric acid	$C_4H_9NO_2$	103.121	103.121
Citrate	$C_6H_5O_7{}^{-3}$	189.099	189.004
Stearic acid	$C_{18}H_{36}O_2$	284.484	284.272
Hexadecanoic acid	$C_{16}H_{32}O_2$	257.422	257.244
Potassium Gluconate	$C_6H_{11}KO_7$	234.245	234.014
Nicotinic acid	$C_6H_5NO_2$	123.111	123.032
Phosphoric acid	H_3PO_4	97.994	97.977
Sodium Pyrophosphate	$Na_4P_2O_7$	265.9	265.871
Acetamide	C_2H_5NO	59.068	59.037
Decanedioic acid	$C_{12}H_{22}O_4$	230.304	230.152
Indoleaceteic acid	$C_{10}H_9NO_2$	175.187	175.063
1-Napthaleneacetic acid	$C_{12}H_{10}O_2$	186.21	186.068

Compound Name	Molecular Formula	Molecular weight (g/ mol)	Exact Mass (g/mol)
4-Hydroxybenzaldehyde	$C_7H_6O_2$	122.123	122.037
2,4-dihydroxybenzaldehyde	$C_7H_6O_3$	138.122	138.032
3,4-Dihydroxybenzoic	$C_7H_6O_4$	154.121	154.027

Table 4.
List of metabolites identified from wild seagrasses.

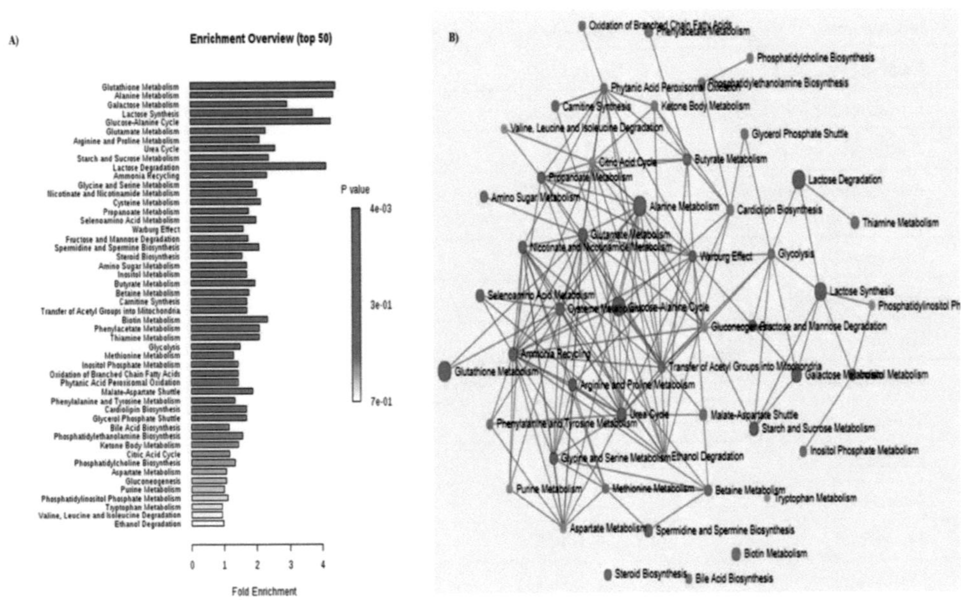

Figure 3.
Distribution of metabolic pathways of differential metabolites derived from wild seagrass *(A) and their active networks (B) (adapted from [53, 54]).*

The sulphated flavonoids have also been traced out in *Halophila ovalis* and *Thalassia testudinum* [72]. It was found that the flavonoid glycosides and acyl derivatives in *P. oceanica* and 15 flavonoid derivatives in *Halophila johnsonii* [73]. Similar to flavonoids and phenolic compounds, the sterol composition has also been reported from temperate seagrasses than the tropical seagrass. The presence of long cahin fatty acids in *P. oceanica* [74] and α-hydroxy fatty acids in *Z. mulleri* [75]. The polar lipids and fatty acids from *Z. marina* and Phyllospadix iwatensis [76], phospholipids and glycolipids from *Z. marina* [77] and steroids, fatty acids from *Z. japonica* [78] confirms the prevalence of volatile compounds in segrasses.

Jeyapragash et al. [53] investigated the systematic identification and characterization of metabolic changes in wild and SCC of *H. pinifolia* using GC–MS based metabolomics approach. It was found that the wild sample accumulated 98 metabolites, while SCC with 125 metabolites along with their relative abundance. The metabolites profiled from wild and SCC was used to map their biochemical pathways. Interestingly, the accumulated metabolites in wild were spanned with 77 primary and secondary metabolism pathways and 73 pathways in SCC. Kyoto Encyclopaedia of Genes and Genomes (KEGG) pathway analysis confirmed insightful biochemical alterations occurred in wild and SCC such as glutathione metabolism, glutamate metabolism, phenylalanine and tyrosine metabolism in wild with significant variations in citric acid metabolism, glycolysis, gluconeogenesis, oxidative phosphorylation related pathways of SCC respectively. Random forest analysis helped to group top 15 metabolites and their correct classification. The data obtained will provide baseline information on *H. pinifolia* metabolism in the marine and artificial environments. Furthermore, it helps to study the stress responsive mechanisms of seagrasses in the marine environment which would further aid in the dissection stress tolerance mechanism.

In addition, Jeyapragash et al. [79] reported the heat stress responsive metabolomics analysis of seagrass *H. pinifolia* in the marine environment. Since, ocean warming is a major global concern in the marine environment, the study focussed to understand the tolerance mechanism of seagrass *Halodule pinifolia* under temperature stress (24, 29, 37, and 45°C) using OMICS approach. Ecophysiologiacal responses such as net photosynthesis (ΔF/F'm) and dark respiration (Fv/Fm) were also studied. Results found that photosynthetic efficiency (ΔF/F'm) significantly reduced due to heat stress, while the rate of dark respiration rate increased as compared to the control (24°C), respectively. Metabolomics study revealed that heat stress could cause huge metabolic alterations with respect to sugar, amino acids and organic acids. Interestingly, thermo-protective compounds such as trehalose (sugar), glycine betaine (amino acid) and methyl vinyl ketone (organic acid) were accumulated from *H. pinifolia* (45°C) which was the first report on the occurrence of glycine betaine and methyl vinyl ketone from seagrasses and other aquatic species so far. These findings would help the research groups to focus on the gene to metabolite networks mechanism for an effective management of seagrass conservation by genetic manipulation.

6. Seagrass cell suspension culture

Plants are considered as the factories of chemical compounds produced in order to carry out their biochemical pathways for survival and propagation [80]. All plants produce secondary metabolites which gained importance in pharmaceutical applications since ancient periods. The plant-based drug discovery gained importance with the development of anti-infectious and anti-cancer drugs which contributes to new bioactive molecules that are being isolated for the treatment of other diseases such as diabetes and obesity [81]. However, the important plant derived drugs are obtained commercially by the extraction from their respective plants. Currently, the natural plant habitats are vanishing due to environmental and geopolitical instabilities and so making it very difficult to procure important secondary metabolites and in the process many potential bioactive compounds have been left undiscovered. Plant cell culture is considered as a promising alternative approach for producing the bioactive compounds that are challenging to be obtained by chemical synthesis or plant extraction [82]. Plant cell culture studies have been carried out on the basis of the totipotent nature, in which the cell has the full set of genes necessary for secondary metabolisms [83]. The production of secondary metabolite via plant tissue culture have been commercialized sincelate 1950s, when atropine from the roots of *Atropa belladonna* was accumulated in roots and callus [84]. The list of metabolites profiled and their metabolic pathways from SCC of seagrasses were shown in **Figure 4** and tabulated in **Table 5**.

Improved plant cell culture techniques made possible to increase the target metabolite production under *in-vitro* conditions. However, agronomically focussed biotechnology is capable to make use of plant tissue culture; such methods are not usually available to marine biologist and notably, *in-vitro* culture of seagrasses has been far more problematic due to lack of the suitable culturing conditions [85].

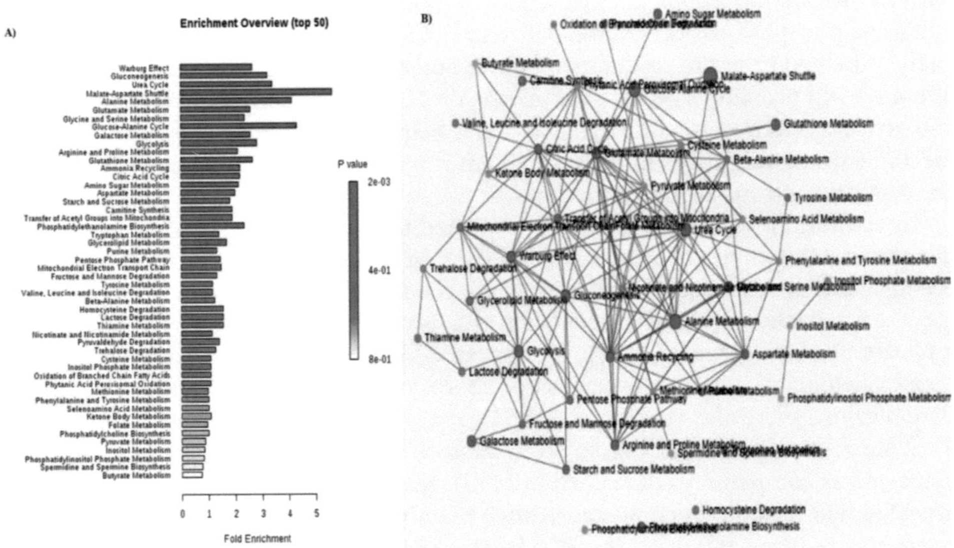

Figure 4.
Distribution of metabolic pathways of differential metabolites derived from SCC of seagrass *(A) and their active networks (B) (adapted from [53, 54]).*

Available literature has brought that 14% of seagrasses are at an elevated risk of extinction under IUCN red list of threatened species [86]. Though numerous studies have been addressed the in-vitro culture of seagrass [38, 87–93], no report exist on the metabolomics analysis from suspension cultured cells of seagrass. *In-vitro* propagation techniques of different seagrass species for restoration and protocol for seagrass protoplast isolation are prevailing [94–98]. Among all these protocols, Carpeneto *et al*. [96] was successful in the cell wall regeneration of protoplast from *Posidonia oceanica* and *Cymodocea nodosa*. Recently, establishment of cell suspension culture has been achieved in *C. nodosa* [99], *Halodule pinifolia, C. rotundata* and *C. serrulata* [38]. Jeyapragash et al., [53, 54] reported that the metabolites synthesized from seagrass *H. pinifolia* in the marine and artificial environment will be highly similar with a total of 98 metabolites in wild and 125 metabolites in SCC respectively and however the cellular suspension accumulated the higher content. The study suggested that the use of seagrass cellular suspension for metabolomics engineering will provide a new facet for novel metabolite identification and characterization. Nevertheless, there lay famine knowledge in the metabolite accumulation pattern and their different biotechnological and pharmaceutical applications. Enhancement of metabolite biosynthesis can also be achieved via the precursors or elicitor treatments of plant cells (**Figure 5**). Precursors are the compound which act as intermediate in or at the beginning of biosynthetic route, treatments using the same stands a good chance of increasing the yield of the desired product. Exogenous supply of biosynthetic precursors to the culture medium induces the high yield of targeted products (Whitmer *et al*., 1998; [100]).

Plant also synthesizes the secondary metabolites to protect themselves in response to various environmental stresses. It might be physical, chemical or a biological factor which induces the higher secondary metabolism known as elicitors. The use of elicitors in cell suspension cultures has been developed to enhance the yield of secondary metabolites, wherein elicitation of target compounds can be induced by the addition of trace number of elicitors [101]. Biotic and abiotic

Query	Match	HMDB	PubChem
Rosmarinic acid	$C_{18}H_{16}O_8$	360.318	360.085
Caffeic acid	$C_9H_8O_4$	180.159	180.042
p-Coumaric acid	$C_9H_8O_3$	164.16	164.047
Protocatacheuic acid	$C_7H_6O_4$	154.121	154.027
p-Anisic acid	$C_8H_8O_3$	152.149	152.047
Vanillic acid	$C_8H_8O_4$	168.148	168.042
Naringenin	$C_{15}H_{12}O_5$	272.256	272.068
4-hydroxybenzoic acid	$C_7H_6O_3$	138.122	138.032
Fructose-6-phosphate	$C_6H_{13}O_9P$	260.135	260.03
Glucose-6-phosphate	$C_6H_{13}O_9P$	260.135	260.03
Glucose	C6H12O6	180.156	180.063
Phosphoenol pyruvic acid	$C_3H_5O_6P$	168.041	167.982
Pyruvic acid	$C_3H_4O_3$	88.062	88.016
Citric acid	$C_6H_5O_7^{-3}$	189.099	189.004
Fumaric acid	$C_4H_4O_4$	116.072	116.011
3-PGA	$C_3H_7O_7P$	186.056	185.993
Ketoglutaric acid	$C_5H_6O_5$	146.098	146.022
Malic acid	$C_4H_6O_5$	134.087	134.022
Succinic acid	$C_4H_6O_4$	118.088	118.027
Mannose	$C_6H_{12}O_6$	180.156	180.063
Oxaloacetic acid	$C_4H_4O_5$	132.071	132.071
Sucrose	$C_{12}H_{22}O_{11}$	342.297	342.116
D-Fructose	$C_6H_{12}O_6$	180.156	180.063
Raffinose	$C_{18}H_{32}O_{16}$	504.438	504.169
Trehalose	$C_{12}H_{22}O_{11}$	342.297	342.116
Turanose	$C_{12}H_{22}O_{11}$	342.297	342.116
Mannitol	$C_6H_{14}O_6$	182.172	182.079
Inositol	$C_{12}H_{22}O_{11}$	342.297	342.116
Xylitol	$C_5H_{12}O_5$	152.146	152.146
Alanine	$C_3H_7NO_2$	89.094	89.048
Aspargine	$C_4H_8N_2O_3$	132.119	132.053
Aspartic acid	$C_4H_7NO_4$	133.103	133.038
Glutamic acid	$C_5H_9NO_4$	147.13	147.053
Glycine	$C_2H_5NO_2$	75.067	75.032
Proline	$C_5H_9NO_2$	115.132	115.063
Serine	$C_3H_7NO_3$	105.093	105.043
Threonine	$C_4H_9NO_3$	119.12	119.058
Valine	$C_5H_{11}NO_2$	117.148	117.079
2,4-dihydroxybenzoic acid	$C_7H_6O_3$	138.122	138.032
2-hydroxybutyric acid	$C_4H_8O_3$	104.105	104.047

Query	Match	HMDB	PubChem
Gamma-aminobutyric acid	$C_4H_9NO_2$	103.121	103.121
Dimethylamine	$(CH_3)_2NH$	45.085	45.058
Ethanolamine	C_2H_7NO	61.084	61.053
Thiamine	$C_{12}H_{17}N_4OS^+$	265.355	265.112
Nicotinic acid	$C_6H_5NO_2$	123.111	123.032
Pyridoxine	$C_8H_{11}NO_3$	169.18	169.074
Phenylanaline	$C_9H_{11}NO_2$	165.192	165.079
Tryrosine	$C_9H_{11}NO_3$	181.191	181.074
Shikimic acid	$C_7H_{10}O_5$	174.152	174.053
Acotinic acid	$C_6H_6O_6$	174.108	174.016
Xylonic acid	$C_5H_{10}O_6$	166.129	166.048
Ascorbic acid	$C_6H_8O_6$	176.124	176.032
Guanine-2′3'-cyclic monophosphate	$C_{10}H_{12}N_5O_7P$	345.208	345.047
Pantothenate	$C_9H_{16}NO_5$	218.229	218.103
Sphingosine	$C_{18}H_{37}NO_2$	299.499	299.28
N-acetylglucosamine	$C_8H_{15}NO_6$	221.209	221.09
Aspartyl leucine	$C_{10}H_{18}N_2O_5$	246.263	246.122
2-hydroxy glutaric acid	$C_5H_8O_5$	148.114	148.037
Glyceric acid	$C_3H_6O_4$	106.077	106.027
Chlorogenic acid	$C_{16}H_{18}O_9$	354.311	354.095
Rhamnose	$C_6H_{12}O_5$	164.157	164.068
Guanosine monophosphate	$C_{10}H_{15}N_5O_{11}P_2$	443.202	443.024
Ribose	$C_5H_{10}O_5$	150.13	150.053
Adenosine-2′3'-cyclic Monophosphate	$C_{10}H_{14}N_5O_7P$	347.224	347.063
Dihydroquercetic acid	$C_{15}H_{12}O_7$	304.254	304.058
Adenosine-2-Monophosphate	$C_{10}H_{14}N_5O_7P$	347.224	347.063
p-hydroxybenzoic acid	$C_7H_6O_3$	138.122	138.032
Quinic acid	$C^7H^{12}O^6$	192.167	192.063
Tryptophan	$C^{11}H^{12}N^2O^2$	204.229	204.09
Pyroglutamic acid	$C_5H_7NO_3$	129.115	129.043
Salicylic acid	$C_7H_6O_3$	138.122	138.032
Methionine	$C_5H_{11}NO_2S$	149.208	149.051
Lactic acid	$C_3H_6O_3$	90.078	90.032
Isovaleric acid	$C^5H^{10}O2$	102.133	102.068
2-oxyglutaric acid	$C_5H_6O_5$	146.098	146.022
2-hydroxyisobutyric acid	$C_4H_8O_3$	104.105	104.047
1-methylnicotinic acid	$C_7H_8NO_2^+$	138.146	138.056
Hypoxanthine	$C_5H_4N_4O$	136.114	136.039

Query	Match	HMDB	PubChem
Indole Acetic acid	$C_{10}H_9NO_2$	175.187	175.063
Naptheline acetic acid	$C_{12}H_{10}O_2$	186.21	186.068

Table 5.
List of metabolites identified from suspension cultured cells seagrasses.

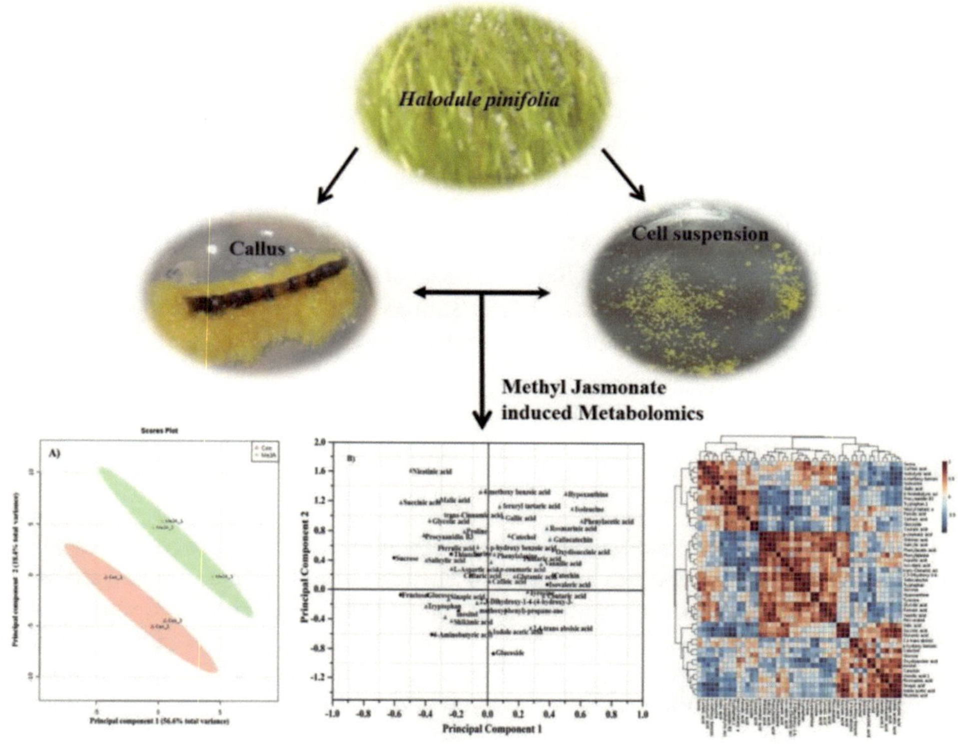

Figure 5.
Elicitor induced comparative metabolomics of wild and cellular suspension of seagrass.

elicitors are available which depends on the target compounds that need to be synthesized.

7. Seagrasses—a source for marine based drug discovery

Ravn *et al.* [102] reported that phenolic acids such as *p*-coumaric acid, caffeic acid, ferulic acid, protocatacheuic acid, *p*-hydroxybenzoic acid, vannillic acid, gentisic acid found predominant in *Halophila ovalis, Thalassia hemprichii, Halodule* sp., *Cymodocea* spp., *Enhalus acoroides, Syriingodium isoetifolium* and other seagrass species. In addition, the pharmaceutically potent rosmarinic acid has been isolated from *Z. marina* [102] *Z. noltii* [103], *H. pinifolia* [51] and also from the detritus of *Z. noltii* and *Z. marina* [57]. Along with rosmarinic acid, caffeic acid and chlorogenic acid found to be present in some trace amounts, since caffeic acid acts as a precursor for roamarinic acid in the shikimic acid biosynthesis. It was also been reported that the predominance of caffeic acid was accumulated in the leaves of *P. oceanica* [104] and *Thalassodendron ciliatum* [105]. Jeyapragash *et al.* [51], profiled 45 metabolites from *Halodule pinifolia*, which includes caffeic acid, coumaric acid, chlorogenic acid and rosmarinic acid which found predominant than other compounds. Recently, biofilm associated, multidrug resistant *Pseudomonas aeruginosa* infection remain a challenging problem in the clinical field since the conventional antibiotic therapy are largely inefficient and new approaches are needed. Inactivating the QS virulence mechanism with anti-infective agent is an attractive approach to prevent bacterial infections without resistance development [54]. Seagrass *Halodule pinifolia* (Miki) Hartog has been shown to exhibit potential antimicrobial activities against *P. aerugonasa* PAO1. Preliminary screening on antibiofilm activity showed that the methanolic extract of *H. pinifolia* exhibited potential inhibition of biofilm formation (96%) as compared to the control respectively. Eight bioactive compounds such as 4-hydroxybenzoic acid, rosmarinic acid, 4-methoxybenzoic acid, *p*-coumaric acid, protocatacheuic acid, caffeic acid, naringenin, vanillic acid, were profiled. Of these compounds, 4-methoxybenzoic acid (4-MBA) showed maximum bacterial growth inhibition that act as a lead molecule with minimum inhibitory concentration (MIC). Furthermore, 4-MBA at MIC concentration reduced the virulence factors and down regulated the level of QS mediated virulence transcripts. The study suggests that seagrasses may act as a newer source for the marine based drug discovery and may act as anti-infective agent against biofilm-mediated harmful pathogens.

8. Conclusion

To summarize, experiments in seagrass metabolomics to date helped us to validate a vast array of metabolites and their alterations in response to various stress mechanisms. This approach has previously enabled to recognize a large number of metabolites whose accumulation is affected upon the exposure of organisms under stress conditions. Nevertheless, despite the many advancements that have been achieved in this field, much work is still needed to identify the seagrass metabolites and their novel metabolic pathways connected to stress response and their tolerance mechanism and to interpret the extensive organization and interaction among gene to metabolite networks. This chapter provides knowledge on the systematic identification and metabolic characterization of seagrass metabolites using metabolomics approach. The bioactive potential of compounds derived from seagrasses paves a way to lead as potential inhibitors of many harmful pathogens in the pharmaceutical sectors and therefore, seagrass explored as newer marine source for the development of plant-based drugs. Further, in-vitro cultures of seagrass afford an alternate model for the up-regulation of enhanced bioactive compound synthesis. Moreover, various stress related metabolomics approach of wild seagrasses should be studied in order to derive diverse group of bioactive metabolites as much as possible, so as to fill the knowledge gap of seagrass metabolites and step forward towards the commercialization of bioactive natural products from seagrasses.

Acknowledgements

The authors would be grateful to University Grants Commission - Basic Scientific Research (UGC-BSR) for their funding and thankful to the Centre of Advanced Study in Marine Biology, Annamalai University for their research facility. Our sincere gratitude to Department of Biotechnology, Karpagam Academy of Higher Education for their scientific interactions to complete the work.

Conflict of Interest

The authors declare that there is no conflict of interest.

Abbreviations

GC–MS	Gas Chromatography–Mass Spectrometry
LC–MS	Liquid Chromatography Mass Spectrometry
FT-ICR-MS	Fourier Transform- Ion Cyclotron Resonance-Mass Spectrometry
SCC	suspension cultured cells
NMR	Nuclear Magnetic Resonance
4-MBA	4-methoxy benzoic acid
MIC	minimum inhibitory concentration
KEGG	Kyoto Encyclopaedia of Genes and Genomes
IUCN	International Union for conservation of Nature
CE-MS	Capillary Electrophoresis-Mass Spectrometry

Author details

Danaraj Jeyapragash[1*], Ayyappan Saravanakumar[2] and Mariasingarayan Yosuva[2]

1 Department of Biotechnology, Karpagam Academy of Higher Education (Deemed to be University), Eachanari, Coimbatore, Tamil Nadu, India

2 Centre of Advanced Study in Marine Biology, Faculty of Marine Sciences, Annamalai University, Parangipettai, Tamil Nadu, India

*Address all correspondence to: pragashdb@gmail.com

References

[1] Pauling L, Robinson AB, Teranishi R, Cary P. Quantitative analysis of urine vapor and breath by gas–liquid partition chromatography. Proceedings of the National Academy of Sciences of the United States of America. 1971;68:2374-2376. DOI: 10.1073/pnas.68.10.2374.

[2] Arabidopsis Genome Initiative. Analysis of the genome sequence of the flowering plant *Arabidopsis thaliana*. Nature. 2000;408:796-815. DOI: 10.1038/35048692

[3] Fowler S, Thomashow MF. Arabidopsis transcriptome profiling indicates that multiple regulatory pathways are activated during cold acclimation in addition to the CBF cold response pathway. Plant Cell. 2002;14 (8):1675-1690. DOI: 10.1105/tpc.003483.

[4] Froehlich JE, Wilkerson CG, Ray WK, McAndrew RS, Osteryoung KW, Gage DA, Phinney BS. Proteomic study ofthe *Arabidopsis thaliana* chloroplastic envelope membrane utilizing alternatives to traditional two-dimensional electrophoresis. Journal of Proteome Research. 2003;2:413-425. DOI: 10.1021/pr034025j.

[5] Hajduch M, Ganapathy A, Stein JW, Thelen JJ. A systematic proteomic study of seed filling in soybean. Establishment of high-resolution two- dimensional reference maps, expression profiles, and an interactive proteome database. Plant Physiology. 2005;137:1397-1419. DOI: 10.1104/pp.104.056614.

[6] Santoni VJ, Vingh J, Pflieger D, Sommerer N, Maurel C. A proteomic study reveals novel insights into the diversity of aquaporin forms expressed in the plasma membrane of plant roots. Biochemical Journal. 2003;373:289-296. DOI: 10.1042/BJ20030159.

[7] Fernie AR, Trethewey RN, Krotzky AJ, Willmitzer L. Metabolite profiling: From diagnostics to systems biology. Nature Reviews Molecular Cell Biology. 2004;5:763- 769. DOI: 10.1038/nrm1451.

[8] Jorge TF, Mata AT, António C. Mass spectrometry as a quantitative tool in plant metabolomics. Philosophical transactions. Philosophical Transactions of the Royal Society A. 2016;374: 20150370. DOI: 10.1098/rsta.2015.0370.

[9] Oksman-Caldentey KM, Inze D. Plant cell factories in the post-genomic era: New ways to produce designer secondary metabolites. Trends in Plant Science.2004;9:433-440. DOI: 10.1016/j.tplants.2004.07.006

[10] Fiehn O, Kopka J, Dörmann P, Altmann T, Trethewey RN, Willmitzer L. Metabolite profiling for plant functional genomics. Nature Biotechnology. 2000; 18(11):1157-1161.

[11] Hall RO, Bernhardt ES, Likens GE. Relating nutrient uptake with transient storage in forested mountain streams. Limnology and Oceanography. 2002;47: 255-265. DOI: 10.4319/lo.2002.47.1.0255.

[12] Seger C, Sturm S. Analytical aspects of plant metabolite profiling platforms: Current standings and future aims. Journal of Proteome Research. 2007;6: 480-497. DOI: 10.1021/pr0604716.

[13] Sumner LW, Lei Z, Nikolau BJ, Saito K. Modern plant metabolomics: Advanced natural product gene discoveries, improved technologies, and future prospects. Natural Product Reports. 2015;32:212-229. DOI: 10.1039/c4np00072b.

[14] Nicholson JK, Lindon JC, Holmes E. Metabonomics: Understanding the metabolic responses of living systems to pathophysiological stimuli via multivariate statistical analysis of biological NMR spectroscopic data.

Xenobiotica.1999;29:1181-1189. DOI: 10.1080/004982599238047.

[15] Olivier DK, Van Wyk BE, Van Heerden FR. The chemotaxonomic and medicinal significance of phenolic acids in *Arctopus* and *Alepidea* (Apiaceae subfamily Saniculoideae). Biochemical Systematics and Ecology. 2008;36:724-729. DOI: 10.1016/j.bse.2008.07.003.

[16] Fiehn O. Combining genomics, metabolome analysis, and biochemical modeling to understand metabolic networks. Comparative and Functional Genomics. 2001;2(3): 155-168. DOI: 10.1002/cfg.82.

[17] Fiehn O. Metabolomics-the link between genotypes and phenotypes. Plant Molecular Biology. 2002;48(1–2): 155-171. DOI:10.1023/A:1013713905833.

[18] Hill CB, Roessner U. Metabolic profiling of plants by GC–MS. The Handbook of Plant Metabolomics (eds. W. Weckwerth and G. Kahl), Wiley; 2013.1-23pp. DOI: 10.1002/9783527669882.ch1.

[19] Hirai MY, Saito K. Post-genomics approaches for the elucidation of plant adaptive mechanisms to Sulphur deficiency. Journal of Experimental Botany. 2004;55(404):1871- 1879. DOI: 10.1093/jxb/erh184.

[20] Roessner U, Willmitzer L, Fernie AR. High-resolution metabolic phenotyping of genetically and environmentally diverse potato tuber systems. Identification of phenocopies. Plant Physiology., 2001;127:749-764. DOI: 10.1104/pp.010316

[21] Soga Y, Ueno Y, Naraoka H, Ohashi Y, Tomita M, Nishioka T. Simultaneous determination of anionic intermediates for *Bacillus subtilis* metabolic pathways by capillary electrophoresis electrospray ionization mass spectrometry. Analytical Chemistry. 2002;74:2233-2239. DOI: 10.1021/ac020064n.

[22] Lindon JC, Holmes E, Nicholson JK. Pattern recognition methods and applications in biomedical magnetic resonance. Progress in Nuclear Magnetic Resonance Spectroscopy. 2001;39:1-40. DOI: 10.1016/S0079-6565(00)00036-4.

[23] Wasim M, Hassan MS, Brereton RG. Evaluation of chemometric methods for determining the number and position of components in high-performance liquid chromatography detected by diode array detector and on flow H-1 nuclear magnetic resonance spectroscopy. Analyst.2003;128:1082-1090. DOI: 10.1039/B304313D.

[24] Meiler J, Will M. Genius: A genetic algorithm for automated structure elucidation from C-13 NMR spectra. Journal of the American Chemical Society.2002;124:1868-1870. DOI: 10.1021/ja0109388.

[25] Halket JM, Zaikin VG. Derivatization in mass spectrometry – 1 Silylation. European Journal of Mass Spectrometry.2003; 9:1-21. DOI: 10.1255/ejms.527.

[26] Sumner LW, Mendes P, Dixon RA. Dixon, plant metabolomics: Large-scale phytochemistry in the functional genomics era. Phytochemistry. 2003;62: 817-836. DOI: 10.1016/s0031-9422(02)00708-2.

[27] Roessner U, Wagner C, Kopka J, Trethewey RN, Willmitzer L. Simultaneous analysis of metabolites in potato tuber by gas chromatography-mass spectrometry. The Plant Journal.2000;23:131-142. DOI: 10.1046/j.1365-313x.2000.00774.x.

[28] Roessner U, Willmitzer L, Fernie AR. Metabolic profiling and biochemical phenotyping of plant systems. Plant Cell Reports.2002;21:189-196. DOI: 10.1007/s00299-002-0510-8.

[29] Aharoni A, de Vos CH, Verhoeven HA, Maliepaard CA, Kruppa G, Bino RJ,

Goodenowe DB. Non-targeted metabolome analysis by use of Fourier transform ion cyclotron mass spectrometry. OMICS A Journal of Integrative Biology. 2002;6(3):217-234. DOI: 10.1089/15362310260256882

[30] Plumb RS, Stumpf CL, Granger JH, Castro-Perez J, Haselden JN, Dear GJ. Use of liquid chromatography/time-of-flight mass spectrometry and multivariate statistical analysis shows promise for the detection of drug metabolites in biological fluids. Rapid Communications in Mass Spectrometry.2003;17:2632-2638. DOI: 10.1002/rcm.1250.

[31] Swart PJ, Bronner GM, Bruins AP, Ensing K, Tepper PG, De Zeeuw RA. HPLC-UV atmospheric-pressure ionization mass-spectrometric determination of the dopamine-D2 agonist N-0923 and its major metabolites after oxidative-metabolism by rat-liver, monkey liver, and human liver Microsomes. Toxicology Methods.1993;3:279-290. DOI: 10.3109/15376519309068445.

[32] Taylor J, King RD, Altmann T, Fiehn O. Application of metabolomics to plant genotype discrimination using statistics and machine learning. Bioinformatics. 2002;18:241- 248. DOI: 10.1093/bioinformatics/18.suppl_2.s241.

[33] Unger M, Dreyer M, Specker S, Laug S, Pelzing M, Neusüss C, Holzgrab U, Bringmann G. Analytical characterisation of crude extracts from an African *Ancistrocladus* species using high-performance liquid chromatography and capillary electrophoresis coupled to ion trap mass spectrometry. Phytochemical Analysis. 2004;15:21-26. DOI: 10.1002/pca.737.

[34] Hall J. Cellular mechanisms for heavy metal detoxification and tolerance. Journal of Experimental Botany. 2002;53(366):1-11. DOI: 10.1093/jexbot/53.366.1.

[35] Knapp DR. Handbook of Analytical Derivatization Reactions. Wiley, One Wiley Drive, Somerset, NJ 08873;1979. 741 pp. DOI: 10.1002/jps.2600690549.

[36] Grob R, Barry E. Modern Practice of Gas Chromatography. Wiley Inter science; 2004. 817 - 818 pp.

[37] Papenbrock J. Highlights in Seagrasses Phylogeny, Physiology, and Metabolism: What Makes them Special? ISRN Botany.2012. 15 p. DOI: 10.5402/2012/103892.

[38] Subhashini P, Dilipan E, Thangaradjou T, Papenbrock J. Bioactive natural products from marine angiosperms: Abundance and functions. Natural Products and Bioprospecting. 2013;3:129-136. DOI: 10.1007/s13659-013-0043-6.

[39] Hartog CD, Kuo J. Taxonomy and biogeography of seagrasses. In: Larkum, A.W.D., Orth, R.J., Duarte, C.M. (Eds.), Seagrasses: Biology, Ecology, and Conservation, Springer, Berlin; 2006.1 - 23pp.

[40] Les DH, Tippery NP. In time and with water the systematics of alismatid monocotyledons. In: Wilkin, P., Mayo, S.J. (Eds.), Early Events in Monocot Evolution. Cambridge University Press, Cambridge. 2013. 118 - 164 pp.

[41] Bremer B, Bremer K, Chase MW, Fay MF, Reveal JL, Soltis DE, Soltis PS, Stevens PF, Anderberg AA, Moore, MJ, Olmstead RG, Rudall PJ, Sytsma KJ, Tank DC, Wurdack K, Xiang JQY, Zmarzty S. An update of the angiosperm phylogeny group classification for the orders and families of flowering plants: APG III. Botanical Journal of the Linnean Society. 2009;161:105-121. DOI: 10.1111/j.1095-8339.2009.00996.x.

[42] Browne CM, Milne R, Griffiths C, Bolton JJ, Anderson RJ. Epiphytic seaweeds and invertebrates associated with south African populations of the

rocky shore seagrass *Thalassodendron leptocaulis* - a hidden wealth of biodiversity. African Journal of Marine Science. 2013;35:523-531. DOI: 10.2989/1814232X.2013.864332.

[43] Duarte MC, Bandeira S, Romeiras MM. Systematics, and ecology of a new species of seagrass (*Thalassodendron*, Cymodoceaceae) from southeast African coasts. Novon A Journal for Botanical Nomenclature. 2012;22:16-24. DOI: 10.3417/2010079

[44] Mabberley DJ. Mabberley Plant-Book, 3rd edition. Cambridge University Press, Cambridge; 2008.

[45] Puglisi MP, Engel S, Jensen PR, Fenical W. Antimicrobial activities of extracts from indo-Pacific marine plants against marine pathogens and saprophytes. Marine Biology. 2007;150: 531-540. DOI: 10.1007/s00227-006-0376-3.

[46] Ross C, Puglisi MP, Paul VJ. Antifungal defenses of seagrasses from the Indian River lagoon. Florida. Aquatic Botany.2008;88:134-141. DOI: 10.1016/j.aquabot.2007.09.003.

[47] Athiperumalsami T, Kumar V, Jesudass LL, Survey and phytochemical analysis of seagrasses in the Gulf of Mannar, southeast coast of India. Botanica Marina. 2008;51: 269-277. DOI: 10.1515/BOT.2008.038.

[48] Kumar CS, Sarada DVL, Gideon TP, Rengasamy R. Antibacterial activity of three south Indian seagrasses, *Cymodocea serrulata*, *Halophila ovalis* and *Zostera capensis*. World Journal of Microbiology and Biotechnology. 2008; 24(9):1989-1992. DOI: 10.1007/s11274-008-9695-5.

[49] Gu J, Weber K, Klemp E, Winters G, Franssen SU, Wienpahl I, Huylmans AK, Zecher K, Reusch TB, Bornberg-Bauer E, Weber AP. Identifying core features of adaptive metabolic mechanisms for chronic heat stress attenuation contributing to systems robustness. Integrative Biology. 2012;4 (5):480-493.doi: 10.1039/c2i b00109h.

[50] Kock WD, Hasler-Sheetal H, Holmer M, Tsapakis M, Apostolaki ET. Metabolomics and traditional indicators unveil stress of a seagrass (*Cymodocea nodosa*) meadow at intermediate distance from a fish farm. Ecological Indicators. 2012; 109:105765. DOI: 10.1016/j.ecolind. 2019.105765

[51] Jeyapragash D, Saravanakumar A, Begum S, Kumar S, Thangaradjou T. GC-MS based metabolomics analysis and occurrence of rosmarinic acid in the tropical seagrass *Halodule pinifolia*. International Journal of Pharmaceutical research. 2018;10(4):277-285. DOI: 10.31838/ijpr/2018.10.04.023.

[52] Gao Y, Jiang Z, Du M, Fang J, Jiang W, Fang J. Photosynthetic and metabolic responses of eelgrass Zostera marina L. to short-term high-temperature exposure. Journal of Oceanology and Limnology. 2018;1-11. DOI: 10.1007/s00343-019-7319-6.

[53] Jeyapragash D, Yosuva M, Karuppiah V. Seagrass *Halodule pinifolia* active constituent 4-methoxybenzioic acid (4-MBA) inhibits quorum sensing mediated biofilm and virulence production of *Pseudomonas aeruginosa*. Microbial Pathogenesis (Elsevier). 2020a;147:104392. DOI: 10.1016/j. micpath.2020.104392.

[54] Jeyapragash D, Yosuva M, Saravanakumar A. Comparative metabolomics analysis of wild and suspension cultured cells of seagrass *Halodule pinifolia*. Aquatic Botany (Elsevier). 2020b;167:103278. DOI: 10.1016/j.aquabot.2020.103278.

[55] Griffiths LL, Melvin SD, Connolly RM, Pearson RM, Brown CJ. Metabolomic indicators for low-light stress in seagrass. Ecological Indicators.

2020; 114:106316. DOI:10.1016/j. ecolind. 2020.106316

[56] Pradheeba M, Dilipan E, Nobi EP, Thangaradjou T, Sivakumar K. Evaluation of seagrasses for their nutritional value. Indian Journal of Geo-Marine Science. 2011;40(1):105-111.

[57] AchamLale S, Rezzonico B, Grignon-Dubois M. Evaluation of *Zostera detritus* as a potential new source of zosteric acid. Journal of Applied Phycology. 2009;21:347- 352. DOI: 10.1007/s10811-008-9375-8

[58] Nuissier G, Diaba F, Dubois MG. Bioactive agents from beach waste: *Syringodium flotsam* evaluation as a new source of L-Chico-inositol. Innovative Food Science and Emerging Technologies.2008;9:396-400. DOI: 10.1016/j.ifset.2007.12.002.

[59] Gillanders B.M.. Seagrasses, fish, and fisheries. In: Larkum, A.W.D., Orth, R.J., Duarte, C.M. (Eds.), Seagrasses: Biology, Ecology, and Conservation. Springer, Berlin; 2006. 503-536 pp.

[60] Gobert S. Cambridge ML, Velimirov B, Pergent G, Lepoint G, Bouquegneau JM, Dauby P, Pergent-Martini C, Walker DI. Biology of *Posidonia*. In: Larkum, A. W. D., Orth, R. J., Duarte, C. M. (Eds.), Seagrasses: Biology, Ecology, and Conservation. Springer, Berlin; 2006. 387-408 pp. DOI: 10.1007/1-4020-2983-7_17.

[61] Ruiz JM, Boudouresque CF, Enriquez S. Mediterranean seagrasses. Botanica Marina. 2009;52:369-381. DOI: 10.1515/BOT.2009.058.

[62] Heglmeier A, Zidorn C. Secondary metabolites of *Posidonia oceanica* (Posidoniaceae). Biochemical Systematics and Ecology. 2010;38:964-970. DOI: 10.1016/j.bse.2010.07.001

[63] Kannan R, Kannan L. Biochemical and caloric composition of seaweeds and seagrass of the Palk Bay, south east coast of India. In: VN Raja Rao, Anand Bishen Singh N, Mahendra Pal Singh Dehradun editors. Algological Research in India: Festchrift. In: , editors; 2002 471 p.

[64] El-Hady HHA, Daboor SM, Ghoniemy AE. Nutritive and antimicrobial profiles of some seagrass from Bardawil Lake, Egypt, Egyptian Journal of Aquatic Research. 2007. 33 (3):103-110.

[65] Arnold TM, Targett NM. Marine tannins: The importance of a mechanistic framework for predictive ecological roles. Journal of Chemical Ecology. 2002;28:1919-1934. DOI: 10.1023/a:1020737609151.

[66] Vergeer LHT, Aarts TL, De Groot JD. The "wasting disease" and the effect of abiotic factors (light intensity, temperature, salinity) and infection with *Labyrinthula zosterae* on the phenolic content of *Zostera marina* shoots. Aquatic Botany.1995;52:35-44. DOI: 10.1016/0304-3770(95)00480-N.

[67] Haznedaroglu MZ, Zeybek U. HPLC determination of chicoric acid in leaves of *Posidonia oceanica*. Pharmaceutical Biology. 2007;45:745-748. DOI: 10.1080/13880200701585717.

[68] Larkum AWD, Drew EA Ralph PJ. Photosynthesis in seagrasses. In: Larkum, A.W.D., Orth, R.J., Duarte, C. M. (Eds.), Seagrasses: Biology, Ecology, and Conservation. Springer, Berlin; 2006.159-192 pp.

[69] Harborne JB, Williams CA. Occurrence of sulphated flavones and caffeic acid esters in members of Fluviales. Biochemical Systematics and Ecology. 1976;4:37-41. DOI: 10.1016/0305-1978(76)90007-7.

[70] Barron D, Varin L, Ibrahim RK, Harborne JB, Williams CA. Sulphated flavonoids, an update. Phytochemistry.

1998; 27: 2375-2395. DOI: 10.1016/0031-9422(88)87003-1,

[71] McMillan C. Flowering under controlled conditions by *cymodocea serrulata*, *halophila stipulacea*, *syringodium isoetifolium*, *zostera capensis* and *Thalassia hemprichii* from Kenya. Aquatic Botany.1980;8:323-336. DOI: 10.1016/0304-3770(80)90062-5.

[72] Rowley DC, Hansen MST, Rhodes D, Sotriffer CA, Ni H, McCammon JA., Bushman FD, Fenical W. Thalassiolins A–C: new marine-derived inhibitors of HIV cDNA integrase. Bioorganic and Medicinal Chemistry. 2002;10(11):3619-3625. DOI:10.1016/s0968-0896(02)00241-9.

[73] Meng Y, Krzysiak AJ, Durako MJ, Kunzelman JI, Wright JLC. Flavones and flavone glycosides from *Halophila johnsonii*. Phytochemistry.2008;69: 2603-2608. DOI:10.1016/j.phytochem.2008.07.007.

[74] Viso AC, Pesando D, Bernard P, Marty JC. Lipid components of the Mediterranean seagrass *Posidonia oceanica*. Phytochemistry. 1993;34:381-387. DOI: 10.1016/0031-9422(93)80012-H.

[75] Volkman JK, Barrett SM, Blackburn SI, Mansour MP, Sikes EL, Gelin F. Microalgal biomarkers: A review of recent research developments. Organic Geochemistry. 1999;29:1163. DOI: 10.1016/S0146-6380(98)00062-X.

[76] Vaskovsky VE, Khotimchenko SV, Xia B, Healthy L, Hefang L. Polar lipids and fatty acids of some marine macrophytes from the yellow sea. Phytochemistry. 1996:42(5):1347- 1356. DOI: 10.1016/0031-9422(96)00117-3.

[77] Goncharova SN, Sanina NM, Kostetsky EY. Role of lipids in molecular thermo adaptation mechanisms of seagrass *Zostera marina*. Biochemical Society Transactions. 2000; 28(6):887-890. DOI: 10.1042/0300-5127:0280887.

[78] Hu H, Wang G, Liu J, Cao H, Zheng X. Studies on phenolic compounds from *Polygonum aviculare*. Zhongguo Zhongyao Za zhi. 2006;31(9):740-742.

[79] Jeyapragash D, Saravanakumar A, Yosuva M, Packiyavathy IASV, Jeba Sweetly D. Chlorophyll fluorescence, dark respiration and metabolomic analysis of *Halodule pinifolia* reveal potential heat responsive metabolites and biochemical pathways under ocean warming. Marine Environmental Research. 2021;164:105248. DOI: 10.1016/j.marenvres.2020.105248.

[80] Zhi-lin Y, Chun-chao D, Lian-quing C. Regulation and accumulation of secondary metabolites in plant-fungus symbiotic system. African Journal of Biotechnology. 2007;6(11):1266-1271.

[81] Saklani A, Kutty SK. Plant-derived compounds in clinical trials. Drug Discovery Today. 2008;13(3–4):161-171. DOI: 10.1016/j.drudis.2007.10.010.

[82] Kloewe ME, Gaurav V, Roberts SC. Pharmaceutically active natural products synthesis and supply vial plant cell culture technology. Molecular Pharmaceutics. 2008;5:243-256. DOI: 10.1021/mp7001494.

[83] Verpoorte R, van der Heijden R, ten Hoopen HJG Memelink J. Metabolic engineering of plant secondary metabolite pathways for the production of fine chemicals. Biotechnology Letters.1999;21:467-479. DOI: 10.1023/A:1005502632053.

[84] West-jr FR, Mika ES. Synthesis of atropine by isolated roots and root-callus cultures of *Belladonna*. Botanical Gazette., 1957:119:50-54. DOI: 10.1086/335961.

[85] Bird KT, Jewett-Smith J. Development of a medium and culture

system for in vitro propagation of the seagrass *Halophila engelmannii*. Canadian Journal of Botany. 1994;72: 1503-1510. DOI: 10.1139/b94-185.

[86] Short FT, Polidoro B, Livingstone SR, Carpenter KE, Bandeira S, Bujang JS, Calumpong HP, Carruthers TJB, Coles RG, Dennison WC, Erftemeijer PLA, Fortes MD, Freeman AS, Jagtap TG, Kamal AHM, Kendrick GA, Kenworthy WJ, Nafie YAL, Zieman, JC. Extinction risk assessment of the world's seagrass species. Biological Conservation. 2011;144(7):1961-1971. DOI: 10.1016/j.biocon.2011.04.010.

[87] Balestri E, Piazzi L, Cinelli F. In vitro germination and seedling development of *Posidonia oceanica*. Aquatic Botany. 1998;60:83-93. DOI: 10.1016/S0304-3770(97)00017-X.

[88] Bird KT, Cody BR, Jewett-smith J, Kane M.E. Salinity effects on *Ruppia maritima* L. cultured in vitro. Botanica Marina.1993;36:23-28. DOI: 10.1515/botm.1993.36.1.23.

[89] Bird KT, Brown MS, Henderson TT, Hara CEO, Robbie JM. Culture studies of Ruppia maritima L. bicarbonate and sucrose-based media. Journal of Experimental Marine Biology and Ecology. 1996;199:153-164. DOI: 10.1016/0022-0981(95)00195-6.

[90] Koch EW, Durako MJ. In vitro studies of the submerged angiosperm *Ruppia maritima*: Auxin and cytokinin effects on plant growth and development. Marine Biology. 1991; 110 (1):1-6. DOI: 10.1007/BF01313085.

[91] Loques F, Caye G, Meinesz A. Axenic culture of selected tissue of *Posidonia oceanica*. Aquatic Botany. 1990;37(2):171-188. DOI: 10.1016/0304-3770(90)90090-8.

[92] Thangaradjou T, Kannan L. Survival and growth of transplants of laboratory raised axenic seedlings of *Enhalus acoroides* (L.f.) Royle and field-collected plants of *Syringodium isoetifolium* (Aschers.) Dandy, *Thalassia hemprichii* (Ehrenb.) Aschers. And *Halodule pinifolia* (Miki) den Hartog. Journal of Coastal Conservation. 2008;12:135-143. DOI: 10.1007/s11852-008-0036-5.

[93] Wilson JG, Bennett IJ. Nutrient requirements of in vitro cultured *Halophila ovalis* and *Posidonia coriacea*: Nitrogen source. Plant Cell, Tissue and Organ Culture.2008;92: 155-163. DOI: 10.1007/s11240-007-9318-0.

[94] Balestri E, Cinelli F. Isolation of protoplast from the seagrass *Posidonia oceanica* (L.) Delile. Aquatic Botany.1992;43:301-304. DOI: 10.1016/0304-3770(92)90074-S.

[95] Balestri E, Cinelli F. Isolation and cell wall regeneration of protoplast from *Posidonia oceanica* and *Cymodocea nodosa*. Aquatic Botany. 2001;70:237-242. DOI: 10.1016/S0304-3770(01)00157-7.

[96] Carpaneto A, Naso A, Paganetto A, Cornara L, Pesce ER, Gambale F. Properties of ion channels in the protoplast of the Mediterranean seagrass *Posidonia oceanica*. Plant, Cell and Environment. 2004;27:279-292. DOI: 10.1111/j.1365-3040.2003.01139.x.

[97] Garril A, Tyerman SD, Findlay GF. Ion channels in the plasma membrane of protoplasts from the halophytic angiosperm *Zostera muelleri*. The Journal of Membrane Biology. 1994;42:381-393. DOI: 10.1007/BF00233443.

[98] Mazzella L, Mauzerall D, Lyman H, Alberte RS. Protoplast isolation and photosynthetic characteristics of *Zostera marina* L. (eel grass). Botanica Marina. 1981;24(5):285- 290. DOI: 10.1515/botm.1981.24.5.285.

[99] Zarranz Elso M, Gonzalez-Henriquez N, Garcia-Jimenez P, Robaina RR. Restoration of *Cymodocea*

nodosa seagrass meadows through seed propagation: Germination in vitro, seedling culture and field transplants. Botanica Marina.2010;53:173-181. DOI: 10.1515/BOT.2010.019

[100] Silvestrini A, Gabriella P, Bruno B, BarbaraM, Robertvan H, Robert V. Effects of alkaloid precursor feeding on a *Camptotheca acuminata* cell line. Plant Physiology and Biochemistry. 2002.40 (9):749-753.DOI:10.1016/S0981-9428 (02)01436-5.

[101] Radman R, Teresa S, Christopher B, Tajalli K. Elicitation of plants and microbial cell systems. Biotechnology and Applied Biochemistry. 2003; 37(1): 91-102. DOI: 10.1042/ba20020118.

[102] Ravn H, Pedersen MF, Borum J, Andary C, Anthoni U, Christophersen C, Nielsen PH. Seasonal variation and distribution of two phenolic compounds, rosmarinic acid and caffeic acid, in leaves and roots-rhizomes of eelgrass (*Zostera marina* L.). Ophelia., 1994;40:51-61. DOI: 10.1080/ 00785326.1994.10429550.

[103] Males Z, Plazibat M. Investigation of chemical composition of *Zostera noltii*. Hornem. Farmaceutski Glasnik. 2000;56(4):109-118.

[104] Dumay O, Costa J, Desjobert JM, Pergent G. Variations in the concentration of phenolic compounds in the seagrass *Posidonia oceanica* under conditions of competition. Phytochemistry. 2004;65:3211-3220. DOI: 10.1016/j. phytochem.2004.09.003.

[105] Hamdy AHA, Mettwally WSA, El Fotouh MA, Rodriguez B, El-Dewany AI, El-Toumy S AA, Hussein MAA. Bioactive phenolic compounds from the Egyptian Red Sea seagrass *Thalassodendron ciliatum*. Zeitschrift für Naturforschung C. 2012;67:291- 296. DOI: 10.1515/znc-2012-5-608.

Chapter 6

Metabolic Profiling of Transgenic Tobacco Plants Synthesizing Bovine Interferon-Gamma

Vladislav V. Yemelyanov, Roman K. Puzanskiy, Mikhail S. Burlakovskiy, Lyudmila A. Lutova and Maria F. Shishova

Abstract

Interferon-gamma belongs to a large family of cytokines – multifunctional secreted proteins involved in animal non-specific immune response. Previously inbred lines of *Nicotiana tabacum* L. plants harboring a heterologous gene of bovine interferon-gamma *Bt-sIFNG* under the control of a constitutive 35S CaMV promoter have been created by *Agrobacterium*-mediated genetic transformation. The antiviral and immunomodulatory activities of plant-produced interferon-gamma in bovine cell culture and laboratory animals (mice) were observed. A state-of-the-art GS-MS technique has been used to identify the possible effect of the transformation on the plant's metabolome. Total profiles included 350 metabolites from leaves, among which 150 substances were identified up to their class and 80 up to the exact metabolite. Metabolite profiling revealed that plants able to synthesize interferon-gamma are characterized by a higher level of amino acids and other substances involved in nitrogen metabolism. In transgenic plants intensification of the secondary metabolism was also detected. Some alterations were distinguished in plant metabolome depending on cultivation conditions.

Keywords: interferon-gamma, *Nicotiana tabacum*, transgenic plant, metabolome

1. Introduction

Metabolic alterations triggered by biotic and abiotic factors in the environment are the basis of a plant development and adaptation. They are closely related to the shifts in primary as well as secondary metabolism that are responsible for synthesis and accumulation of different regulatory and defensive metabolites [1–3]. Metabolic profiling is a useful tool for distinguishing these alterations. Powerful approaches to provide metabolomic investigations are based on recent technology such as GC–MS (gas chromatography–mass spectrometry), LC–MS (liquid chromatography–MS), CE–MS (capillary electrophoresis–MS), and FI-ICR–MS (Fourier transform ion cyclotron resonance–MS), combined with mass spectrometry (MS) and NMR (nuclear magnetic resonance spectroscopy) [4–6]. Detection of metabolites is followed by multivariate statistical analysis of the accumulated datasets. Qualitative and quantitative changes in metabolic profiles during development and under stress conditions are analyzed by metabolomics, a discipline known as a part of systems biology.

Aside from native metabolic alterations, metabolome profiling is used to detect the changes in plant metabolism brought about by genetic engineering. Such improvements in metabolites serve to elevate the economic importance and environmental sustainability of agricultural plants by increasing herbicide tolerance and resistance to pests and pathogens. Plants possess a high biochemical potential of synthesizing an enormous number of various natural substances: fatty acids, phenolics, terpenoids, alkaloids, glucosinolates and other biologically active compounds; many of which are of great value to pharmacology [7]. Advances in metabolomics have enabled the decoding of many metabolic networks and provide for the active genetic engineering of pathways. These offer the opportunity to manipulate the biosynthesis of valuable, biologically active substances of interest, and to create "design" biochemicals. They also can serve to considerably alter the secondary metabolism thus improving a plant's capacities to synthesize new substances or, on the contrary, reduce their toxicity by switching off such metabolic pathways [8–10]. This is a direct effect of genetic engineering on the metabolome. Nevertheless, activation of one key enzyme will not always result in facilitation of the whole metabolic pathway. Thus metabolic profiling is an important tool to evaluate results of such genetic modification. The technology of genetic engineering is widely used in modern agriculture. Frequently plants are genetically engineered to improve their productivity or tolerance to adverse environment. Among the modified crops are maize, cotton, soybean, canola, rice, tomato, potato, among others [2, 11–13]. Other promising approaches are the use of biotech plant systems as a platform for the production of various heterologous proteins for the pharmaceutical industry [7, 14, 15]. The major advantage of plants as bioreactors, aside from their autotrophic type of nutrition, is that the edible plant with the recombinant protein of interest can be eaten directly, skipping rather expensive procedure of purifying the target protein. The production of immunogenic fragments of recombinant antigens in a bioreactor plant is called "plant-based vaccine". Apart from vaccines, biotech plants are used to synthesize numerous antibodies, cytokines, hormones, and other proteins [7, 14–16].

Social, ethical, economic, and ecological norms demand that biotech crops be subjected to intense scrutiny [17]. A very important question is whether accumulation of a large amount of "foreign protein" could be considered a stress factor by the engineered plant and be accompanied by an accumulation of compounds that would pose an ancillary potential risk. A series of investigations were focused on the evaluation of differences in the metabolic profiles of transgenic plants in comparison with wild types (WT). The main conclusion is that metabolism after transgenesis suffered only insignificant alterations or about the same in comparison to that of WT plants [12]. Moreover, results indicate that environmental variations usually produce greater major differences in metabolome composition than genetic modifications. To uncover a possible problem, we used inbred lines of *Nicotiana tabacum* L. plants synthesizing bovine interferon-gamma to study the effects of genetic transformation on the plant metabolome. It must be noted that interferon-gamma is a powerful stimulator of the immune system against pathogens and the tumors of various origins and has no enzymatic activity. Therefore its possible effect on metabolism would be indirect.

2. Material and methods

2.1 Plant material

In this study two separate inbred lines of tobacco harboring a heterologous gene of bovine interferon-gamma *Bt-sIFNG* under the control of a constitutive 35S

CaMV promoter were used. These lines were created earlier by the *Agrobacterium*-mediated genetic transformation of *Nicotiana tabacum* L., cv. Trabzon [18]. The transformation generated six independent kanamycin-resistant transgenic plants (T_0), of which only two became the founders of homozygous inbred lines: InterB.6 and Inter311.2 (T_1). They demonstrated stable inheritance and expression of the transgene insertion and presence of a biologically active heterologous interferon-gamma protein in plant tissues [18]. T_4 generation of transgenic plants, InterB.6.13.8-1 and Inter311.2.7.2-1, and WT tobacco (as a control) was used for the study. The Inter311.2.7.2-1 plant and WT were grown *in vivo* in soil culture in a greenhouse at a temperature of 22–25 °C with a 16 h photoperiod. InterB.6.13.8-4 and WT were cultivated *in vitro* in a MS0 medium [19] supplemented with sucrose (20 g/L) in sterile conditions under the same temperature and light conditions.

2.2 RNA extraction

Total plant RNA was extracted from tobacco leaves with a Pure-ZOL™ reagent according to the manufacturer's protocol (Bio-Rad Laboratories, Hercules, CA, USA). Samples were then treated with DNase I (Thermo Fisher Scientific, Waltham, MA, USA) (5 U/sample). After that the RNA was precipitated with ethanol to remove residual DNase I. Purified RNA was dissolved in sterile water and stored at −80 °C until analysis.

2.3 Reverse transcription

Two μg of total RNA were taken for cDNA synthesis using MMLV RT kit (Evrogen, Moscow, Russia) in accordance with manufacturer's protocol. cDNA samples were precipitated with 0.1 M sodium acetate in ethanol, diluted with sterile deionized water, aliquoted and stored at −80 °C until analysis.

2.4 PCR analysis

The PCR reaction mixture included Taq-polymerase (Evrogen), a mixture of 2 μM dNTP, magnesium-containing Taq-buffer, cDNA template (100–200 ng), the forward and reverse primers sIFNG-1 (5'-AGGAGTATGGACATCATCAAGCA-3′) and sIFNG-2 (5'-AGTCGTCGACCGGAATTTGA-3′) for *Bt-sIFNG* (product size 105 b.p.), and EF-1 (5'-CAAGCGGTCATTCAAGTATGC-3′) and EF-2 (5'-TGTCCAGGACGATCAATCACA-3′) for tobacco *Nt-EF-1α* gene (product size 135 b.p.) [20]. The tobacco house-keeping gene was used as a control present in both WT and transgenic plants. The primers were ordered from Evrogen. Amplification of the fragments corresponding to *sIFNG* and *Nt-EF-1α* genes was performed using the following program:

95°C, 2 min —
{95°C, 20 s,
60°C, 30 s,
72°C, 30 s} — 32 cycles,
72°C, 1 min
in CFX96 Touch Deep Well Real-Time PCR Detection System (Bio-Rad Laboratories).

2.5 Agarose gel electrophoresis

Electrophoresis was performed in a 1% agarose gel in a TAE buffer with the addition of ethidium bromide (0.5 μg/mL) for 40 min at a voltage of 120 V. PCR fragments were visualized in UV light.

2.6 Sample preparation for metabolome analysis

Average samples of 3–5 leaves were prepared in four biological replicates. The plant material (0.2 g) was frozen in microtubes with liquid nitrogen and ground three times in a Tissue Lyser LT (Quiagen, Düsseldorf, Germany) bead mill with metal balls 5 mm in diameter (50 hits/s, 2 min) and subjected to a single-stage extraction with two mL of methanol. After the centrifugation at 15,000 g for 15 min at 4 °C the supernatant was collected and evaporated in a refrigerated CentriVap centrifugal concentrator (Labconco, Kansas City, MO, USA) at 10 °C. The dried residue was dissolved in pyridine with tricosane (nC23) as an internal standard. The samples were then supplied with the silylating agent BSTFA: TMCS 99:1 (Sigma-Aldrich, St. Louis, MO, USA) and derivatizated at 90 °C for 20 min [21, 22].

2.7 Gas chromatography and mass spectrometry (GC–MS)

An Agilent 5860 gas chromatograph (Agilent Technologies, Santa Clara, CA, USA) with Agilent 7893 autosampler, and Agilent ChemStation E.02.02.1431 software were used for analysis. The samples were separated on a J&W HP-5MS capillary column 30 m long, 0.25 mm in diameter, stationary phase film (95% dimethylpolyoxane, 5% diphenyl), thickness 0.1 mm. The helium gas constant flow was 1 mL/min and the inlet temperature was 250 °C. The temperature parameters of the oven included an initial temperature of 70 °C and a linear increase to 320 °C at the rate of 4 °C/min. Chromatograms were registered with an Agilent 5975C mass selective detector. Metabolite profiling was performed using equipment of the Center for Molecular and Cell Technologies of Research park of St. Petersburg State University.

2.8 GC/MS data processing

GC–MS data was processed using the PARADISe program (Department of Food Science Faculty of Science, University of Copenhagen, Denmark, [23]) coupled with NIST MS Search (National Institute of Standards and Technology (NIST), USA). In addition, the AMDIS (Automated Mass Spectral Deconvolution and Identification System, NIST, USA) was used. The following mass-spectrometer libraries were applied: NIST2010, library of the Resource Center of the Science Park "Center for Molecular and Cell Technologies" (St. Petersburg State University), the Golm Metabolome Database (GMD) and MoNA (Massbank of North America). Retention index (RI) was determined by calibration with standard alkanes.

2.9 Data analysis and visualization

Data analysis was performed in the environment of the R language 3.6.3 "Holding the Windsock" [24]. Data were normalized by internal standard (nC23) as well as by sample median. The data were log-transformed and standardized. Metabolite that was not detected but present in other replicated samples was considered a technical error and missing values were imputed. Missing data imputation was performed by KNN (k-nearest neighbors) with an *impute* R package [25]. The heatmap was made with ComplexHeatmap [26]. PCA (Principal Component Analysis) was realized with *pcaMethods* [27]. OPLS-DA was performed in the *ropls* package [28].

For enrichment analysis, the *fgsea* package was used [29]. For the statistical ranking factor, loadings of the predictive components from OPLS-DA models were used. Pathways associated with this metabolite set were extracted from KEGG with the *KEGGREST* package using *Nicotiana tabacum* as a reference organism [30]. The resulting lists of metabolites for pathways were manually curated. Metabolites identified just up to class were included in the pathways related to these groups. Pathways were mapped within the Cytoscape software environment [31]. Graph nodes were assigned to KEGG pathways, edges represent the presence of common metabolites.

3. Results and discussion

Previous studies generated two different families of transgenic tobacco harboring the heterologous gene of bovine interferon-gamma *Bt-sIFNG* [18, 32]. One homozygous *Bt-sIFNG* plant was identified in the Inter311.2 family at T_1 generation. It established the founder of the 311.2 inbred line of transgenic *Bt-sIFNG* tobacco. The other was obtained in T_2 generation of the InterB.6.13 family, leading to a second inbred line of tobacco synthesizing interferon-gamma. In the current study, plants from the T_4 generation of transgenic lines, Inter311.2.7.2-1 and InterB.6.13.8-1, as well as wild type (WT) tobacco for control were used. To demonstrate the transgenic nature of 311.2.7 and B.6.13, RNA was extracted and reverse transcribed. RT-PCR was performed with primers for the tobacco gene *Nt-EF-1α*, encoding housekeeping protein elongation factor 1α, and for the *Bt-sIFNG* gene, encoding bovine interferon-gamma. It was clear that PCR products corresponding to *Nt-EF-1α* were observed for cDNAs from all tested plants, both WT and transgenic. RT-PCR products with primers for *Bt-sIFNG* gene were found in transgenic plants only (**Figure 1**).

Therefore plants of transgenic inbred lines used in the study were shown to possess and express *Bt-sIFNG* gene insertion. These plants synthesize heterologous interferon-gamma protein at about 1 to 1.5 μg per g of fresh weight of plant tissue [18]. The biological effects of plant-produced bovine interferon-gamma demonstrated antiviral activity in bovine cell culture and the induction of immune response in mice, manifested in qualitative alteration of peripheral blood lymphocytes and overall level of Ig G [18].

The production of recombinant human interferon-gamma expressed in *Escherichia coli* resulted in accumulation of 0.35 g/g of dry weight [33]. Similarly *E. coli* culture produced human interferon-gamma with a speed 3 mg/L in hour [34]. Apart from bacteria, different yeast systems based on transformed lines of *Pichia pastoris* and *Saccharomyces cerevisiae* are used for interferon production of up to 300 mg/L [34]. Several examples using insect or even mammalian cell lines have also been designed. Thus in spite of all the economic costs for production and further purification, all of are characterized by rather intensive protein synthesis.

Another interferon-producing system on the basis of rice suspension culture was shown to accumulate intracellular human interferon-gamma protein of up to

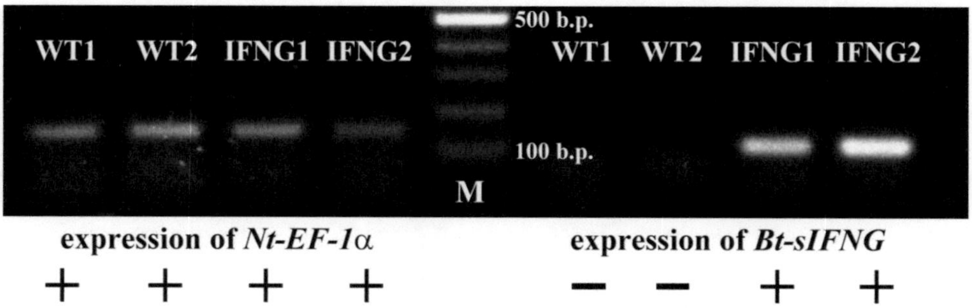

Figure 1.
Electrophoresis of RT-PCR products of cDNAs obtained from WT plant grown in a pot (WT1), or in vitro (WT2), transgenic 311.2.7.2-1 plant, grown in a pot (IFNG1), and transgenic B6.13.8-1 plant, cultivated in vitro (IFNG2), with primers for tobacco gene Nt-EF-1α, or heterologous gene Bt-sIFNG. M - DNA ladder (100–500 b.p.).

699.79 ng/g of cells [35]. The most effective genetic construction used was starvation-inducible endogenous rice αAmy3 promoter.

An example of intensive synthesis of functionally active endogenous plant protein in leaf cells is the accumulation of the major photosynthetic enzyme – RUBISCO (D-Ribulose-1,5-bisphosphate carboxylase/oxygenase), which is claimed to be the most abundant plant protein on earth, even if recently the exact value of the carboxylase has come under revision [36]. The amount of the RUBISCO protein ranged from 30 to 50% of total soluble protein in leaf cell.

Synthesis of protein due to heterologous expression might seriously affect amino acid balance and/or cause alterations in carbohydrates as sources of energy. But along with that, an unregulated accumulation of "foreign" protein inside a plant cell could become a stress signal itself which would lead to production of defensive metabolites. For example, in microbial-based bioreactor systems overproduction of recombinant protein led to formation of an insoluble protein bodies in cytosol and development of oxidative stress as was shown in *E. coli* [37], *P. pastoris* and *S. cerevisiae* [38]. Thus it is appropriate to assume that the tobacco plant metabolism might be shifted due to transgenesis. Earlier comprehensive metabolic analysis of transformed plants showed different alterations in various biochemical processes, nevertheless there was not a clearly established link between gene manipulation and changes in metabolism [12]. The aim of this work was to distinguish possible rearrangements of tobacco leaf cells' metabolic profiles due to transformation with a heterologous gene of bovine interferon-gamma *Bt-sIFNG* under the control of a constitutive 35S CaMV promoter. It was decided that additional attention should be focused on two types of cultivation (in pot soil culture and *in vitro* sterile conditions) – both suitable for transgenic plants growing to estimates of their effect on metabolic alterations.

Metabolite profiles of tobacco leaves were performed by GS-MS analysis. In total profiles included 350 metabolites (**Figure 2**). 80 of these were identified up to the exact metabolite and 150 were identified up to the metabolite class. Metabolic profiles were characterized by a wide variety of carbohydrates and their derivatives (about 90), including pentoses, hexoses and oligosaccharides and their derivatives such as sugar alcohols and sugar acids. This pattern is typical for profiling plant metabolites. The profiles were completed with 19 amino acids, including proteinogenic ones; about 20 carboxylic acids, mainly energy metabolism intermediates; and quite a few fatty acids and their derivatives (only 8). Moreover, some secondary metabolites were also identified.

Simple unsupervised dimension reduction methods (PCA, **Figure 3**) showed differences in tobacco leaf metabolome due to both the type of growth and to genetic status. Metabolite profiles are visualized in the score space of the first two principal components (PC). Metabolomes were grouped along to PC1 according to the type of growth (35% dispersion) and were grouped along with PC2 accordingly to the absence/presence of *Bt-sIFNG* gene insertion (28% dispersion). Similar metabolite profile grouping was also characterized by a correlation of metabolite content. Spearman's correlation coefficient was used as a measure of distance (1-r) and multidimensional scaling (MDS) to reduce the dimension (**Figure 4**). The obtained results highlighted the importance of both factors, namely the method of plant cultivation and its transgenic status.

Further comparison of metabolite profiles of WT and Inter311.2.7.2-1 transgenic tobacco plants grown in pots was detailed by supervised methods such as PLS-DA. The predictive component of PLS-DA has a 31% rate of dispersion ($R2Y = 0.99$, $Q2Y = 0.83$). So the metabolic shifts prompted by the transgenic construct were significant. A bar plot of factor loadings of the predictive component from OPLS-DA is presented in **Figure 5**. Positive values correspond to higher content in the WT. According to results, leaves of the WT tobacco plant contain a higher

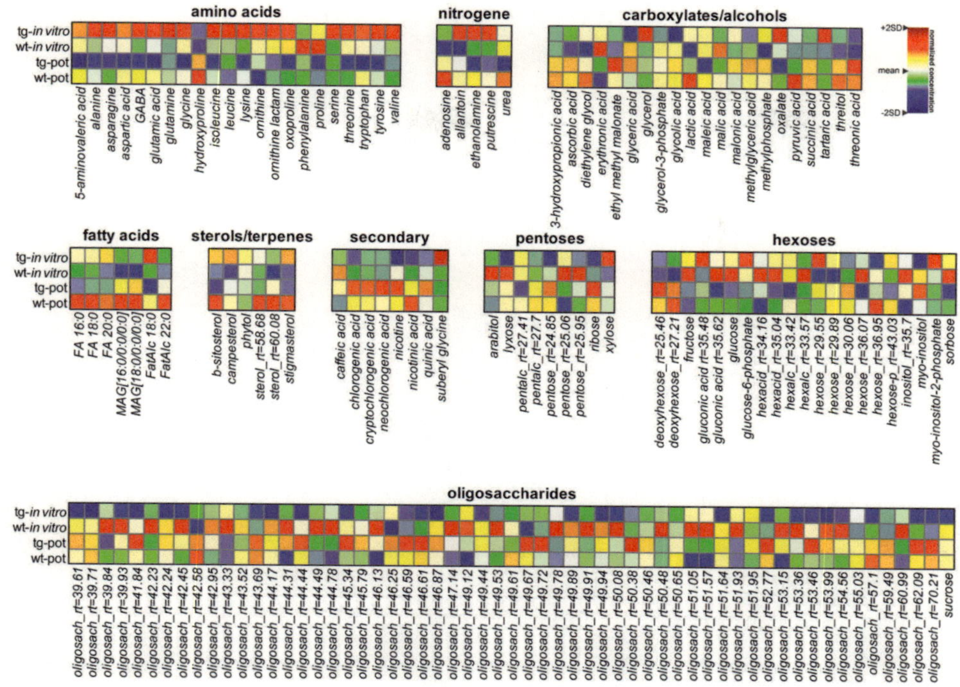

Figure 2.
Heatmap of metabolite content in WT and transgenic (tg) tobacco plants grown in pot or in vitro culture (logarithmic, normalized to median and standardized values).

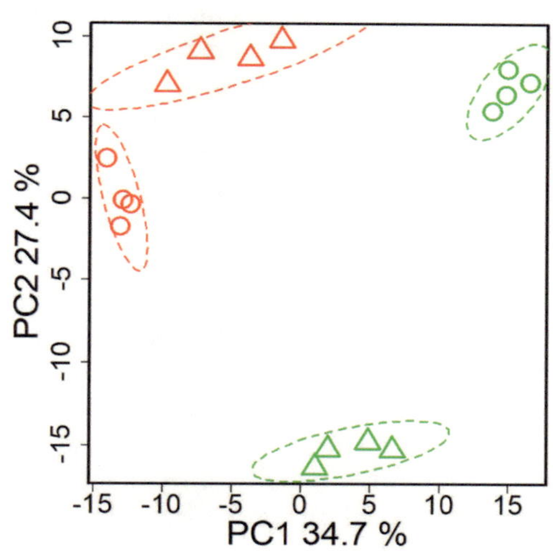

Figure 3.
Representation of metabolite profiles of WT (Δ) and transgenic (o) tobacco in low-dimensional spaces. PCA score plot, % - percent of variance, ellipses - 90% CI. red - pot cultivation, green - in vitro cultivation.

amount of lipophilic compounds, including acylglycerols, sterols, and some fatty acids. WT was also characterized by a higher content of amino acids, amines and carboxylates. Leaves of transgenic plants accumulated high levels of oligosaccharides. Subsequent enrichment analysis revealed metabolic processes that occurred in WT and transgenic tobacco plants (**Figure 6**). In agreement with the bar plot, the results of enrichment analysis of WT plants indicated an intensive wide spectrum of metabolic pathways responsive for the balance of amino acids, carboxylates, lipophilic metabolites, and others. Furthermore, biotech tobacco was characterized

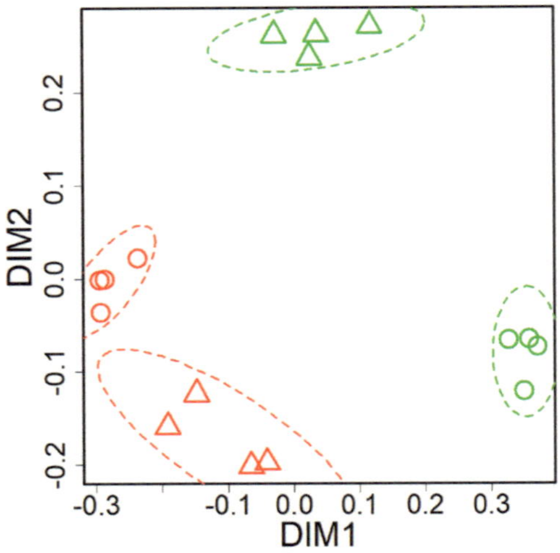

Figure 4.
Representation of metabolite profiles in low-dimensional spaces. Metabolite profiles in the space revealed using MDS with 1-rho as a measure of distance, where rho is Spearman's correlation coefficient, ellipses - 90% CI. Δ - wt, o - transgenic, red - pot culture, green - in vitro cultivation.

by an intensification of only the carbohydrate metabolism pathways. Thus the procedure of growing a plant in soil pot culture exerted significant effects on its metabolism. Such cultivation activated different aspects of WT plant metabolism involving the developmental activity of young leaves. Surprisingly, the presence of an interferon-gamma synthesis-encoding heterologous construct intensified carbohydrate metabolism. The possible higher amount of amino acids required for excessive synthesis of heterologous protein were not distinguished, perhaps because of the intensification of protein synthesis itself.

Another pair of plants were grown in sterile conditions with the application of *in vitro* technology that provides a full supply of required nutrients. Therefore the next step was a similar comparative analysis of metabolite profiles of WT and InterB.6.13.8-1 transgenic plants cultivated *in vitro*. The OPLS-DA provided showed that 65% of the dispersion was associated with the predictive component (R2Y = 1.0, Q2Y = 0.98). Thus the influence of the genotype is more pronounced under *in vitro* conditions. The bar plot of factor loadings of the predictive component from OPLS-DA is presented in **Figures 2** and 7. Leaves of the WT plant accumulated different sugars, oligosaccharides, a number of amino acids, and lipophilic compounds. The transgenic plant, in contrast to pot cultivation, was characterized by a higher diversity of metabolite classes. Metabolic profiles consist of a large number of amino acids, carboxylates, sterols (stigmasterol, β-sitosterol and campesterol), as well as other lipophilic compounds. Nonetheless, enrichment analysis proved activation of only those pathways related to amino acids in transgenic plants and pathways of sugar metabolism in WT plants (**Figure 8**).

Taken together data obtained for WT and transgenic plants grown in pots and *in vitro* were used for SUS-plot construction (**Figure 9**). Unexpectedly, the effect of transformation was the opposite at different conditions of plant cultivation. It might be assumed that the rich conditions of *in vitro* growth cause more intensive metabolic alterations due to genetic modification. These more "comfortable" conditions for plant development might enhance the intensification of heterologous gene expression and interferon-gamma synthesis. This supposition is indirectly confirmed by higher accumulation of PCR-products of the *Bt-sIFNG* transcript in the leaves of transgenic tobacco grown *in vitro* (**Figure 1**).

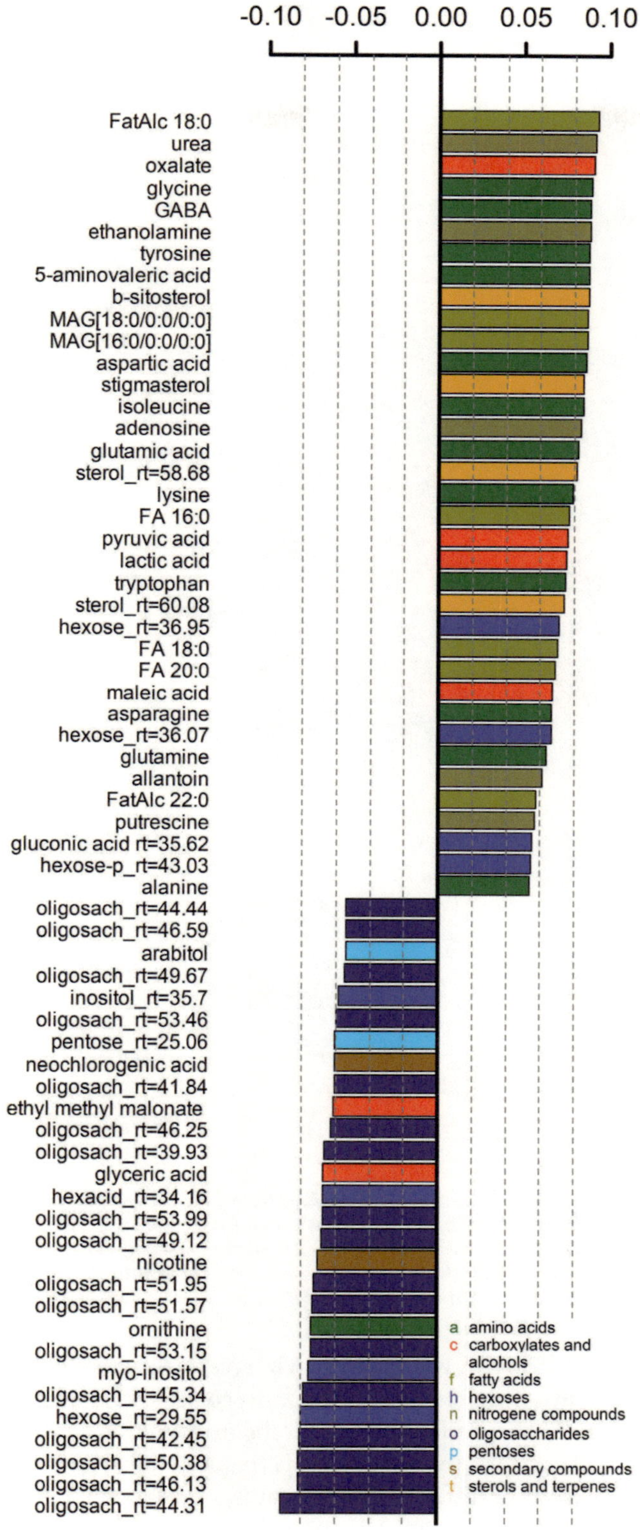

Figure 5.
Differences between metabolomes of WT and transgenic plants cultivated in pots. Barplot of factor loadings of the predictive component from OPLS-DA. Positive values correspond to a higher content in the wild type. Colors mirror chemical class (legend same as in **Figures 7** *and* **9***). Abbreviations: rt. - retention time, oligosacch -oligosaccharides, FA - fatty acid, FatAlc - fatty alcohols, hexacid, pentacid, hexalc, pentalc - sugar acids and alcohols, MAG - monoacylglycerols.*

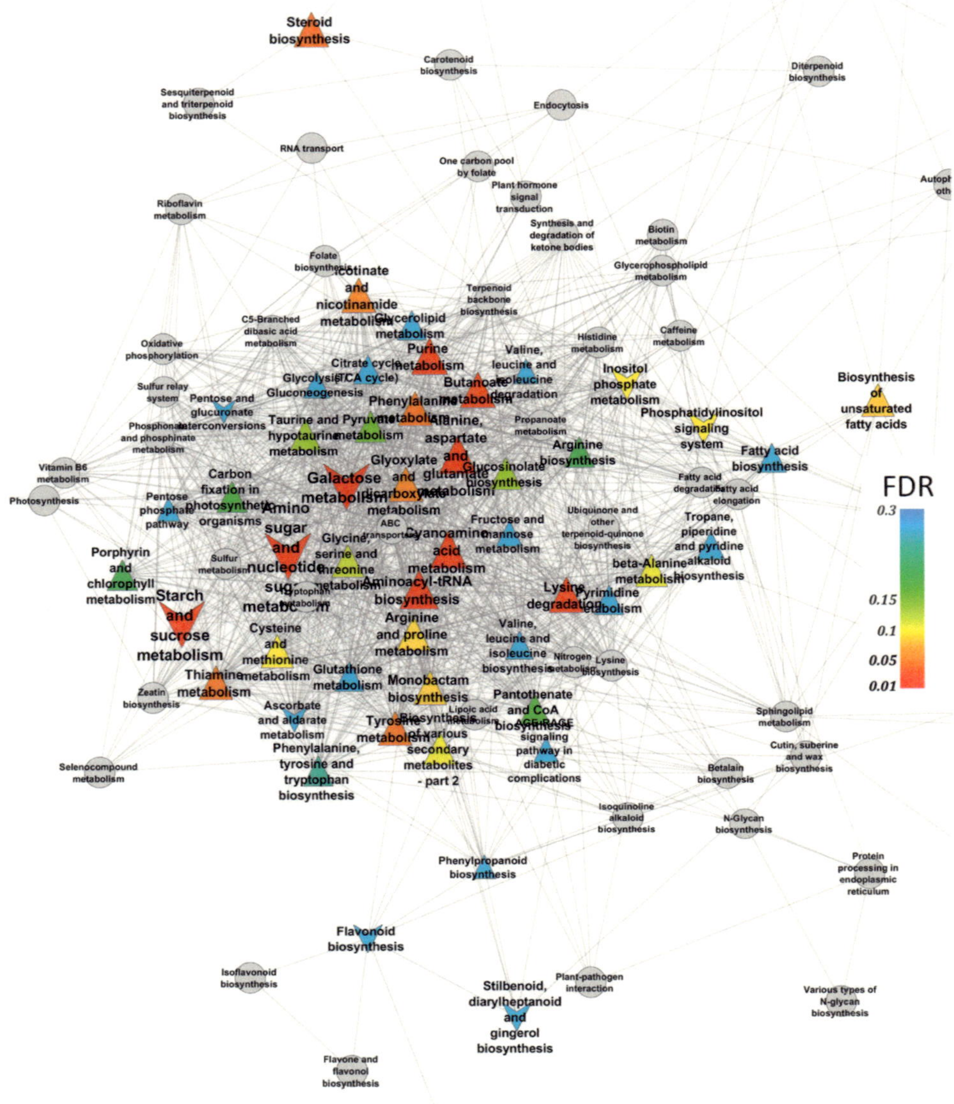

Figure 6.

Enrichment analysis based on predictive component loadings from OPLS-DA model for WT and transgenic tobacco plants cultivated in pots. The network of biochemical pathways (nodes) of tobacco, if the paths have common metabolites, then they are connected by edges. Colors mark significance of enrichment (FDR - false discovery rate), size reflects level of enrichment (|NES| - normalized enrichment score), upward direction - activation (NES > 0) in WT, downward - repression (NES <0) in WT.

An additional criterion that would be of an interest is the crosslink between different metabolic pathways. Thus one can expect some correlative alterations in metabolite content. Changes in the correlations of the metabolite content may reflect systemic metabolic changes. Therefore, we examined the frequency distribution of the correlation coefficients in plants of different genetic status, grown under different conditions. Based on the analysis of the frequency distribution of the Pearson correlation coefficient (**Figure 10**), it was determined that in transgenic plants the number of strong correlations increases when cultivated in a pot in comparison with *in vitro* conditions. The observed effect was absent in WT plants. This might reflect the more stressful cultivation conditions for transgenic plants in pot culture. It may be assumed that the increase in correlations is associated with some kind of coordinated response to stress.

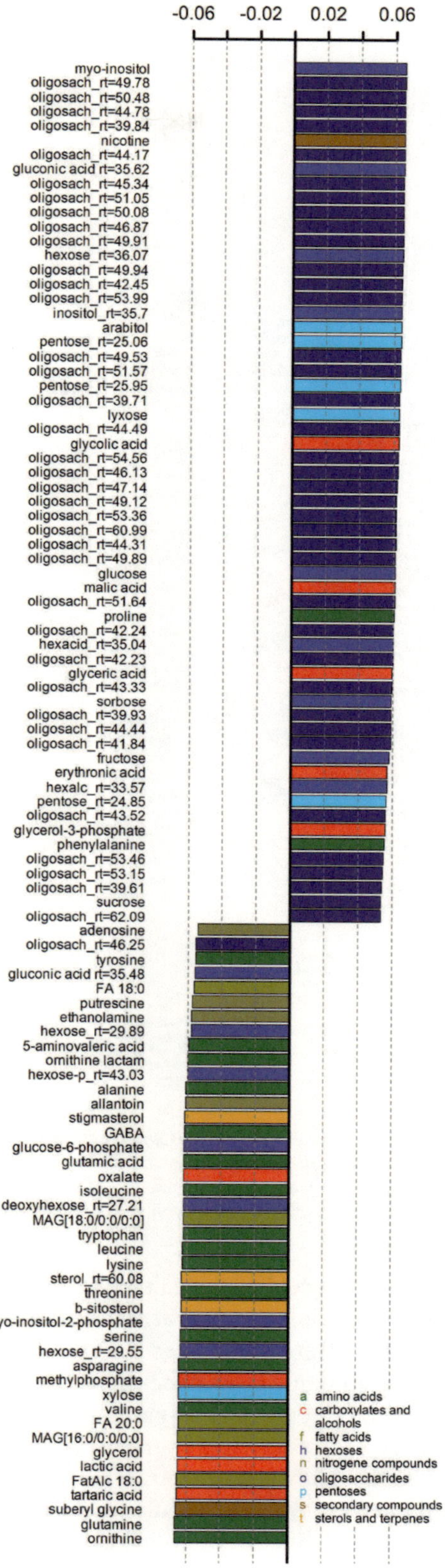

Figure 7.
Differences between metabolomes of WT and transgenic plants cultivated in vitro. Barplot of factor loadings of the predictive component from OPLS-DA. Positive values correspond to a higher content in the wild type. Colors mirror chemical class. Legend and abbreviations as in **Figure 5**.

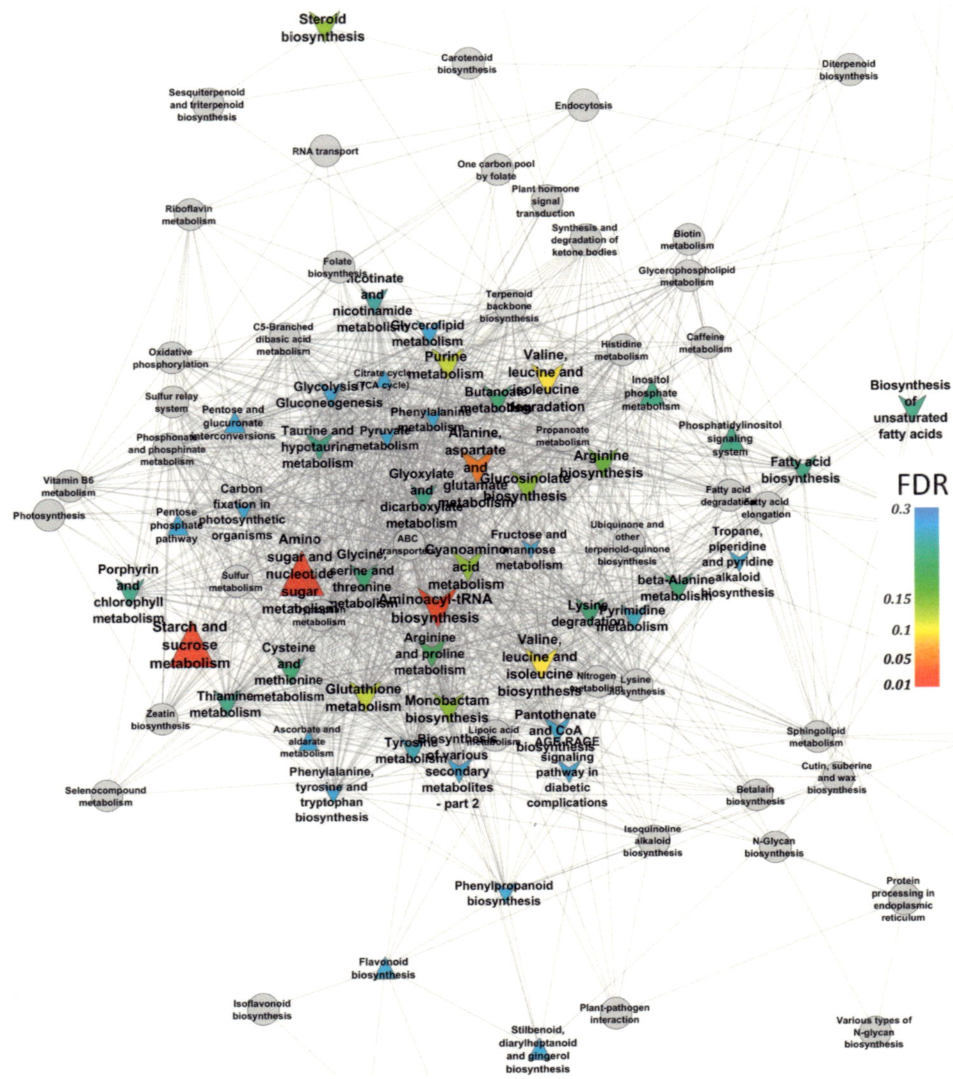

Figure 8.

Enrichment analysis based on predictive component loadings from OPLS-DA model for WT and transgenic plants cultivated in vitro. The network of biochemical pathways (nodes) of tobacco, if the paths have common metabolites, then they are connected by edges. Colors mark significance of enrichment (FDR), size reflects level of enrichment (|NES|), upward direction - activation (NES > 0) in WT, downward - repression (NES <0) in WT.

Analysis of selected literature sources revealed that the majority of work on genetically-modified plant metabolome was carried out with plants expressing heterologous genes encoding different enzymes [12]; i.e. these types of transgenes directly affect the metabolome by enhancing pathways operated by the target enzyme protein. The bovine interferon-gamma used in this study belongs to cytokines and has no enzymatic activity. Therefore, its effects on metabolism would be indirect. There is no doubt that introduction of excessively synthesized "foreign" protein would interfere with the cellular metabolism at least at the level of consumption of an additional amount of amino acids and energy (i.e., sugars and carboxylates). The results obtained showed accumulation of oligosaccharides in the transgenic plant Inter311.2.7.2-1 (compared to WT), grown in soil pot culture (**Figures 2** and **5**) and stimulation of amino acid and sterol metabolism in InterB.6.13.8-1 cultivated *in vitro* in Murashige-Skoog medium (**Figures 2** and 7).

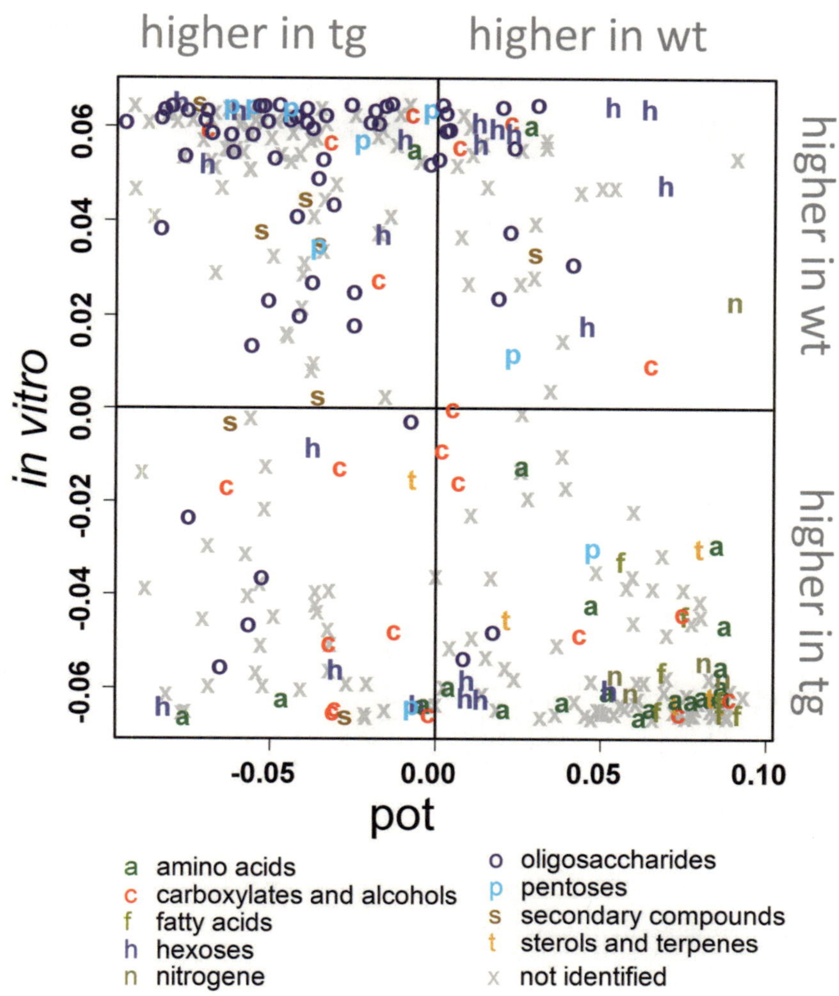

Figure 9.
Comparative analysis of the transformation effects in vitro and in the pot culture. SUS-plot of predictive loadings of OPLS-DA models for WT and transgenic (tg) plants comparison during pot and in vitro cultivation. Colored letters represent chemical classes.

We did not succeed in finding any report on effects of heterologous interferon-gamma on the metabolome of its bioreactor organism, neither of plant, nor of animal origin. Exogenously applied human interferon-gamma led to accumulation of some amino acids (Ala, Tyr, etc.) and depletion of pyruvate and lactate in the culture of human mesenchymal stem cells [39]. Despite the fact that tobacco is a non-target organism for bovine interferon-gamma, its synthesis in InterB.6.13.8–1 grown *in vitro* also led to significant accumulation of amino acids (Lys, GABA, Trp, Asp, Asn, Ser, Glu, Gln, Gly, Tyr, Ile, Leu, Val, Ala, Thr and oxoproline); levels of lactic and pyruvic acids were more or less the same in transgenic plants as in those compared with WT (**Figures 2** and 7).

There are quite a few studies of the metabolome of transgenic plants synthesizing non-enzymatic heterologous protein which can be compared with the system producing bovine interferon-gamma and they mostly concern different types of Bt-toxin expressing plants [12, 13, 40]. It was shown that both cultivation conditions and gene modification induced similar alterations of metabolomes. Moreover, in many cases the growing conditions or developmental stage of plants had a greater effect on the metabolome than the presence of a transgenic insert.

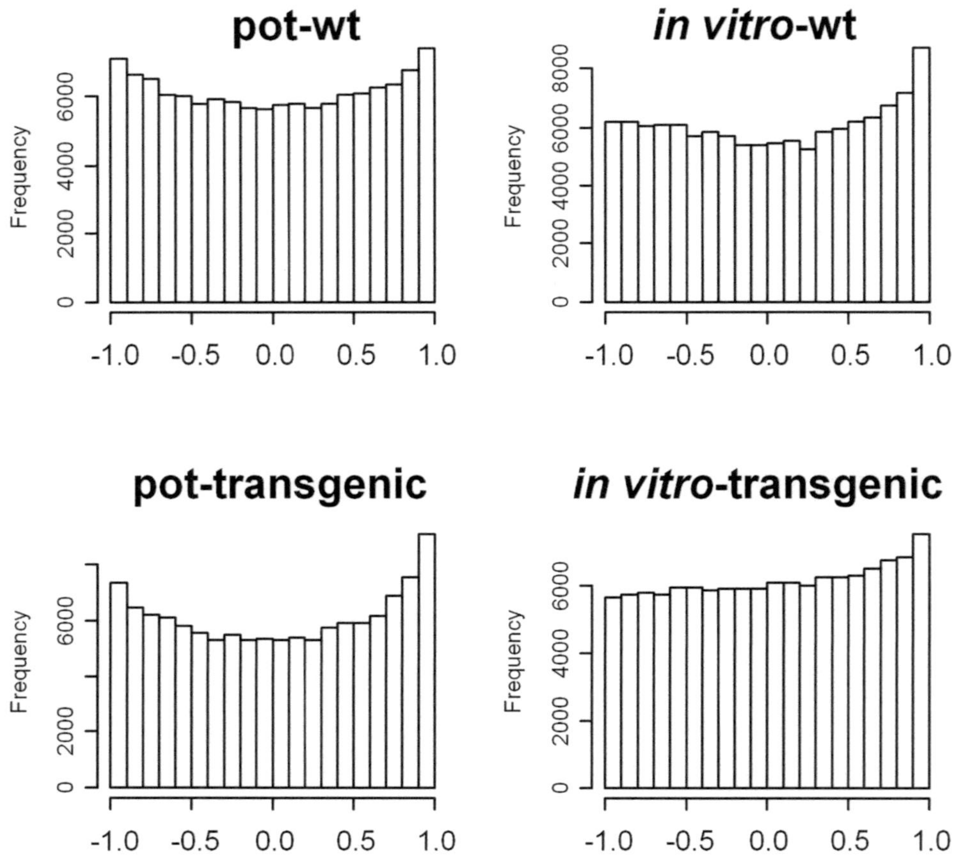

Figure 10.
Influence of transformation and cultivation method on the correlation of metabolite content. Histograms of frequencies of Pearson's correlation coefficient.

So further investigations of biotech plant metabolomics are strongly required, especially those focused on the metabolomics of bioreactor plants.

4. Conclusion

Taken together, the data of this investigation clearly showed that metabolic profiles are dynamic parameters which characterize plant development. Metabolic profile changes in tobacco plants specialized in the synthesis of bovine interferon-gamma were discovered. The presence of heterologous of *Bt-sIFNG* gene insertion itself did not cause development of intracellular stress in transgenic tobacco. In fact, even more intensive metabolic alterations were determined to be dependent on the type of plant growing conditions. It was assumed that soil conditions might be considered as somewhat stressful for transgenic bioreactor plants in comparison with cultivation *in vitro*. This could cause depletion of the target gene transcript accumulation and thus indirectly indicate active regulation of "foreign" protein synthesis by the host transgenic plant. The results obtained are in accordance with data in the literature illustrating the priority of growing conditions and developmental stage above transgenic status in the determination of metabolic intensity. Nevertheless, it is too early in the resolution of this question to make firm conclusions and further investigation is still needed.

Acknowledgements

Research was performed at the Center for Molecular and Cell Technologies of Research park of St. Petersburg State University. The chapter was written under the support of the Ministry of Science and Higher Education of the Russian Federation in accordance with agreement № 075-15-2020-922, dated 16.11.2020 providing a grant in the form of subsidies from the Federal budget of the Russian Federation. The grant was provided for state support for the creation and development of the World-class Scientific Center "Agrotechnologies for the Future".

Conflict of interest

The authors declare no conflict of interest.

Abbreviation

FDR	false discovery rate
GC–MS	gas chromatography–mass spectrometry
MDS	multidimensional scaling
NES	normalized enrichment score
OPLS-DA	orthogonal partial least squares discriminant analysis
PC	principal component
PCA	principal component analysis
PCR	polymerase chain reaction
PLS-DA	partial least squares discriminant analysis
RI	retention index
SUS	shared and unique structures
T_X generation	generation of transgenic plants
T_0	transformed plants
T_1	first generation, obtained after self-pollination of T_0 plants
T_2	second generation, obtained after self-pollination of T_1 plants, etc.
WT	wild type

Author details

Vladislav V. Yemelyanov[1], Roman K. Puzanskiy[2,3], Mikhail S. Burlakovskiy[1], Lyudmila A. Lutova[1] and Maria F. Shishova[2*]

1 Department of Genetics and Biotechnology, St. Petersburg State University, St. Petersburg, Russia

2 Department of Plant Physiology and Biochemistry, St. Petersburg State University, St. Petersburg, Russia

3 Laboratory of Dynamics of the Arctic Vegetation Cover, Komarov Botanical Institute of the Russian Academy of Sciences, St. Petersburg, Russia

*Address all correspondence to: mshishova@mail.ru

References

[1] Wolfender J-L, Rudaz S, Choi YH, Kim HK. Plant metabolomics: From holistic data to relevant biomarkers. Current Medicinal Chemistry. 2013;20:1056-1090. DOI: 10.2174/0929867311320080009

[2] Hong J, Yang L, Zhang D, Shi J. Plant metabolomics: An indispensable system biology tool for plant science. International Journal of Molecular Sciences. 2016;17:767. DOI: 10.3390/ijms17060767

[3] Datir SS, Jha RM. Plant metabolomics. In: Egbuna C, Ifemeje JC, Udedi SC, Kumar S, editors. Phytochemistry: Fundamentals, Modern Techniques, and Applications. 1st ed. CRC Press, Imprint Apple Academic Press, 2018. 22 p. eBook DOI: 10.1201/9780429426223

[4] Okazaki Y, Saito K. Recent advances of metabolomics in plant biotechnology. Plant Biotechnology Reports. 2012;6:1-15. DOI: 10.1007/s11816-011-0191-2

[5] Khakimov B, Bak S, Engelsen SB. High-throughput cereal metabolomics: Current analytical technologies, challenges and perspectives. Journal of Cereal Science. 2014;59:393-418. DOI: 10.1016/j.jcs.2013.10.002

[6] Kumar R, Bohra A, Pandey AK, Pandey MK, Kumar A.Metabolomics for plant improvement: Status and prospects. Frontiers in Plant Science. 2017;8:1302. DOI: 10.3389/fpls.2017.01302

[7] Saveleva NV, Burlakovskiy MS, Yemelyanov VV, Lutova LA. Transgenic plants as bioreactors to produce substances for medical and veterinary uses. Russian Journal of Genetics: Applied Research. 2016;6:712-724. DOI: 10.1134/S2079059716060071

[8] Oksman-Caldentey K-M, Inze D. Plant cell factories in the post-genomic era: New ways to produce designer secondary metabolites. Trends in Plant Science. 2004;9:433-440. DOI: 10.1016/j.tplants.2004.07.006

[9] Staniek A, Bouwmeester H, Fraser PD, Kayser O, Martens S, Tissier A, van der Krol S, Wessjohann L, Warzecha H. Natural products – modifying metabolite pathways in plants. Biotechnology Journal. 2013;8:1159-1171. DOI: 10.1002/biot.201300224

[10] Niazian M. Application of genetics and biotechnology for improving medicinal plants. Planta. 2019;249:953-973. DOI: 10.1007/s00425-019-03099-1

[11] Stitt M, Sonnewald U. Regulation of metabolism in transgenic plants. Annual Review of Plant Physiology and Plant Molecular Biology. 1995:46:341-368. DOI: 10.1146/annurev.pp.46.060195.002013

[12] Simo C, Ibáñez C, Valdés A, Cifuentes A, García-Cañas V. Metabolomics of genetically modified crops. International Journal of Molecular Sciences. 2014;15:18941-18966. DOI: 10.3390/ijms151018941

[13] Zhang X, Zhang R, Li L, Yang Y, Ding Y, Guan H, Wang X,·Zhang A, Wen H. Negligible transcriptome and metabolome alterations in RNAi insecticidal maize against *Monolepta hieroglyphica*. Plant Cell Reports. 2020;39:1539-1547. DOI: 10.1007/s00299-020-02582-4

[14] Sharma AK, Sharma MK. Plants as bioreactors: Recent developments and emerging opportunities. Biotechnology Advances. 2009;27:811-832. DOI: 10.1016/j.biotechadv.2009.06.004

[15] Moon K-B, Park J-S, Park Y-I, Song I-J, Lee H-J, Cho HS, Jeon J-H, Kim H-S. Development of systems for the production of plant-derived

biopharmaceuticals. Plants. 2020;9:30. DOI: 10.3390/plants9010030

[16] Burlakovskiy MS, Yemelyanov VV, Lutova LA. Plant based bioreactors of recombinant cytokines (Review). Applied Biochemistry and Microbiology. 2016;52:121-137. DOI: 10.1134/S0003683816020034

[17] Schiemann J, Dietz-Pfeilstetter A, Hartung F, Kohl C, Romeis J, Sprink T Risk assessment and regulation of plants modified by modern biotechniques: current status and future challenges. Annual Review of Plant Biology. 2019;70:699-726. DOI: 10.1146/annurev-arplant-050718-100025

[18] Burlakovskiy MS, Saveleva NV, Yemelyanov VV, Padkina MV, Lutova LA. Production of bovine interferon-gamma in transgenic tobacco plants. Plant Cell, Tissue and Organ Culture. 2015;22:685-697. DOI: 10.1007/s11240-015-0802-7

[19] Murashige T, Skoog F. A revised medium for rapid growth and bio assays with tobacco tissue cultures. Physiologia Plantarum. 1962;15:473-497. DOI: 10.1111/j.1399-3054.1962.tb08052.x

[20] Guo Y, Li R, Asgher Z, Shamsi IH, Yu S, Zhao J, Ren X. Genome-wide selection and testing of superior reference genes for quantitative gene expression normalization in tobacco (*Nicotiana tabacum*). Research & Reviews: Journal of Microbiology and Biotechnology. 2016;5:68-75.

[21] Puzanskiy RK, Shavarda AL, Tarakhovskaya ER, Shishova MF. Analysis of metabolic profile of *Chlamydomonas reinhardtii* cultivated under autotrophic conditions. Applied Biochemistry and Microbiology. 2015;51:83-94. DOI: 10.1134/S0003683815010135

[22] Puzanskiy RK, Yemelyanov VV, Kliukova MS, Shavarda AL, Shtark OY, Yurkov AP, Shishova MF. Optimization of metabolite profiling for black medick (*Medicago lupulina*) and peas (*Pisum sativum*). Applied Biochemistry and Microbiology 2018;54:442-448. DOI: 10.1134/S0003683818040129

[23] Johnsen L, Skou P, Khakimov B, Bro R. Gas chromatography – mass spectrometry data processing made easy. Journal of Chromatography A. 2017;1503:57-64. DOI: 10.1016/j.chroma.2017.04.052

[24] R Core Team. 2020. R: a language and environment for statistical computing. Vienna, Austria: R Foundation for Statistical Computing. Available at https://www.R-project. org/ [Accessed: 2020-12-27]

[25] Hastie T, Tibshirani R, Narasimhan B, Chu G. *impute: impute:* Imputation for microarray data. R package version 1.64.0. 2020. https://bioconductor.org/packages/release/bioc/html/impute.html [Accessed: 2020-12-28]

[26] Gu Z, Eils R, Schlesner M. Complex heatmaps reveal patterns and correlations in multidimensional genomic data. Bioinformatics. 2016;32:2847-2849. DOI: 10.1093/bioinformatics/btw313

[27] Stacklies W, Redestig H, Scholz M, Walther D, Selbig J. *pcaMethods* a bioconductor package providing PCA methods for incomplete data. Bioinformatics. 2007;23:1164-1167. DOI: 10.1093/bioinformatics/btm069

[28] Thevenot E, Roux A, Xu Y, Ezan E, Junot C. Analysis of the human adult urinary metabolome variations with age, body mass index, and gender by implementing a comprehensive workflow for univariate and OPLS statistical analyses. Journal of Proteome Research. 2015;14:3322-3335. DOI: 10.1021/acs.jproteome.5b00354

[29] Korotkevich G, Sukhov V, Sergushichev A. Fast gene set enrichment

analysis. bioRxiv. 2019. DOI: 10.1101/060012

[30] Tenenbaum D, Maintainer B. *KEGGREST: Client-side REST access to the Kyoto Encyclopedia of Genes and Genomes (KEGG)*. R package version 1.30.1. 2020. DOI: 10.18129/B9.bioc.KEGGREST

[31] Shannon P, Markiel A, Ozier O, Baliga NS, Wang JT, Ramage D, Amin N, Schwikowski B, Ideker T. Cytoscape: A software environment for integrated models of biomolecular interaction networks. Genome Research. 2003;13: 2498-2504. DOI: 10.1101/gr.1239303

[32] Saveleva NV, Lutova LA. Bioreactor Plants for Synthesis of Recombinant Proteins of Medicinal Use. Production of Bovine γ-interferon. Saarbrücken: LAPLAMBERT Academic Publishing; 2010. 160 p. [In Russian]

[33] Khalilzadeh R, Shojaosadati SA, Maghsoudi N, Mohammadian-Mosaabadi J, Mohammadi MR, Bahrami A, Maleksabet N, Nassiri-Khalilli MA, Ebrahimi M, Naderimanesh H. Process development for production of recombinant human interferon-γ expressed in *Escherichia coli*. Journal of Industrial Microbiology and Biotechnology. 2004;31:63-69. DOI: 10.1007/s10295-004-0117-x

[34] Razaghi A, Owens L, Heimann K. Review of the recombinant human interferon gamma as an immuno-therapeutic: Impacts of production platforms and glycosylation. Journal of Biotechnology. 2016;240:48-60. DOI: 10.1016/j.jbiotec.2016.10.022.

[35] Chen T-L, Lin Y-L, Lee Y-L, Yang N-S, Chan M-T. Expression of bioactive human interferon-gamma in transgenic rice cell suspension cultures. Transgenic Research. 2004;13:499-510. DOI: 10.1007/s11248-004-2376-8

[36] Bar-Ona YM, Milo R. The global mass and average rate of RUBISCO. Proceedings of the National Academy of Sciences of USA. 2019;116:4738-4743; DOI: 10.1073/pnas.1816654116

[37] Jurgen B, Lin HY, Riemschneider S, Scharf C, Neubauer P, Schmid R, Hecker M, Schweder T. Monitoring of genes that respond to overproduction of an insoluble recombinant protein in *Escherichia coli* glucose-limited fed-batch fermentations. Biotechnology and Bioengineering. 2000;70:217-224. DOI: 10.1002/1097-0290(20001020)70: 2<217::AID-BIT11>3.0.CO;2-W

[38] Gradoboeva AE, Padkina MV. The heterologous protein production impact on the physiological state of *Saccharomyces cerevisiae* and *Pichia pastoris* yeast. Biological Communications. 2008;53:56-61. [In Russian]

[39] Wobma HM, Tamargo MA, Goeta S, Brown LM, Duran-Struuck R, Vunjak-Novakovic G. The influence of hypoxia and IFN-γ on the proteome and metabolome of therapeutic mesenchymal stem cells. Biomaterials. 2018;167:226-234. DOI: 10.1016/j.biomaterials.2018.03.027

[40] Zhou J, Ma C, Xu H, Yuan K, Lu X, Zhu Z, Wu Y, Xu G. Metabolic profiling of transgenic rice with *cryIAc* and *sck* genes: An evaluation of unintended effects at metabolic level by using GC-FID and GC-MS, Journal of Chromatography B. 2009;877:725-732. DOI: 10.1016/j.jchromb.2009.01.040

Chapter 7

Metabolomic Changes in Wood Inhabiting Filamentous Fungi during Ontogenesis

Katerina V. Sazanova, Nadezhda V. Psurtseva and Alexey L. Shavarda

Abstract

GC–MS-based metabolomic profiling of different strains of basidiomycetes *Lignomyces vetlinianus, Daedaleopsis tricolor* and *Sparassis crispa* were studied. On different stages of growth in the methanol extracts of mycelium the different compounds including amino acids, organic acid of TCA cycle, sugars, fatty acids, sugar alcohols, and sugar acids were detected. Changes in the metabolite network occurring with age of the mycelium of *L. vetlinianus* and *D. tricolor* are discussed. The exponential phase of mycelium growth is characterized by pronounced differences during of growth, which manifests itself both in the analysis of specific compounds and in the modeling of the statistical model of the metabolic network. The metabolomic network in the stationary growth phase is less susceptible to changes over time, and is also characterized by a lower dispersion of samples from one aging group. For some compounds, including biotechnologically significant ones, targeted analysis by GC–MS was performed. 4, 6-dimethoxy-phthalide (4, 6-dimetoxy-1 (3H) -isobenzofuranone) was isolated from the mycelium of *Lignomyces vetlinianus*, accumulating in the mycelium in the form of large aggregates. The accumulation of sparassol and other orsellinic acid derivatives in *Sparassis crispa* culture under various conditions is described.

Keywords: metabolomic profiling, metabolite network, gas chromatography–mass spectrometry, wood inhabiting fungi, ontogenesis, phthalide

1. Introduction

The interactions and interconversions of small biomolecules that form the metabolic network play an important, if not decisive, role in the development and ontogenetic processes of living organisms [1]. It is well known that in addition to the wide variability of the content of each metabolite in any biological object in the course of its vital activity, the transition from one stage of ontogenesis to the next is associated with a number of systemic changes in the qualitative and quantitative composition of metabolites and their ratios [2, 3], i.e. "Metabolic state" (metabolome). Thus, in the last decade in physiological research, the study of the dynamics of the metabolome of the system with the help of metabolic profiling taking into account the time factor has become increasingly important. The results of research in the field of metabolomics largely depend on the instrumental approach, detection, methods of quantitative assessment and localization of metabolites [4]. To implement this approach, it is necessary to analyze complex mixtures using methods such as NMR (nuclear magnetic resonance), GC–MS (gas chromatography–mass spectrometry), LC–MS (liquid chromatography-mass spectrometry), CE-MS (capillary electrophoresis-mass spectrometry) [5, 6]. The application of a modern analytical platform in biology has two main results: metabolomic profiling and the dynamics of metabolic networks. The general purpose of metabolomic profiling is to document the set of metabolites [7]. This task has an independent meaning, since characterizes the molecular resources of the system, i.e. its potential in the context of the concept of biodiversity. For physiology, of particular interest is not so much the annotated metabolome as the dynamics of its correlation structure in the process of development and vital activity. Our laboratory has long been using classical methods of multivariate statistics ("-omic" approaches) with a categorical load in the form of a time scale.

Metabolome analysis represents a tool that finds common application in all aspects investigations, and hence it represents a focal point in studies of fungal physiology [8]. Filamentous fungi have a very active and diverse metabolism resulting in the production of a broad range of metabolites. Fungal chemodiversity is important source of the new compounds for biotechnology and medicine. This makes fungi very promising and interesting object for metabolomic studies, both for the study of the fundamental regularities of the metabolic network, and for the targeted analysis of biotechnologically significant compounds. Below a brief introductions of basidial fungi being an object of this investigation is presented.

Common in Russia bracket fungus *Daedaleopsis tricolor* (Bull.) Bondartsev & Singer (*Polyporaceae*, *Polyporales*, *Agaricomycetes*) is often considered as a color variation of *D. confragosa* (Bolton) J. Schröt. It causes a white rot of dead deciduous and coniferous trees (**Figure 1**). It is known as medicinal fungus producing bioactive compounds with antibiotic [9], antiviral [10], antihypertensive [11] and anti-tumor [12] effects.

Figure 1.
Fruiting bodies of Daedaleopsis tricolor *in nature.*

Pleurotoid xylotrophic fungus originally have been described from Poland as *Pleurotus vetlinianus* Domański but in 2008 was found in Russia (**Figure 2**). Its taxonomy was settled as *Lignomyces vetlinianus* (Domański) R.H. Petersen &

Figure 2.
Fruiting bodies of Lignomyces vetlinianus *in nature.*

Figure 3.
Fruiting bodies of Sparassis crispa *in nature.*

Zmitr. (Resupinataceae, Agaricales, Agaricomycetes), and culture characters were studied [13]. It was noticed then that some strains growing in Petri plates produced agglomerates of light colored crystals formed on colony mats, which were later identified as phthalides [14]. Since this species is not studied well, its physiological and biochemical properties are of quite interest.

Xylobiontic aphyllophoroid basidiomycete *Sparassis crispa* (Wulfen) Fr. (*Sparassidaceae*, *Polyporales*, *Agaricomycetes*) grows in the southern part of the Holarctic, parasitizing on the roots of coniferous trees (**Figure 3**). It belongs to rare species. *S. crispa* has the ability to produce various biologically active substances. Many compounds synthesized by *S. crispa* are well studied, as a rule, these are substances with therapeutic properties [15, 16]. Nevertheless, the physiological and biochemical characteristics of *S. crispa*, which determine its ecological specificity, have not been sufficiently studied, which makes it possible to consider this species as an interesting object for ecological and physiological studies.

The aim of this study is to investigate the dynamic changes in the metabolomic profile of presented species of fungi during their cultures growth using GC–MS analysis.

2. Materials and methods

2.1 Cultivation and cultural studies

The following strains from the Komarov Botanical Institute Basidiomycetes Culture Collection were used for the study: *Daedaleopsis tricolor* LE-BIN 2266, *Sparassis crispa* LE-BIN 043, LE-BIN 2902, LE-BIN 2967 and *Lignomyces vetlinianus* LE-BIN 2335, LE-BIN 2339, LE-BIN 3253.

For culture characters, the strains were cultivated in Petri plates 90 or 60 mm in diam. on beer-wort agar (BWA) (4% beer-wort, Severnie pivovarni, Russia; 2% agar, Difco) at 25 °C. Linear growth was measured every other day for *D. tricolor* and every week for *S. crispa* and *L. vetlinianus* strains (n = 4). Macromorphology of *D. tricolor* was described at 6 weeks, of *S. crispa* and *L. vetlinianus* – at 8 weeks. Micromorphology was studied under Zeiss Axio Imager A1, Axio Scope A1 and Stemi 2000-CS (Zeiss, Germany) using transmitted light.

For metabolomic analysis the strains were also grown in Petri plates: *S. crispa* and *D. tricolor* on BWA, whereas *L. vetlinianus* on malt extract agar (1,5% MEA, Oxoid, England) at 25 °C. Mycelium samples for the analysis of *D. tricolor* were taken on the 3rd, 7th, 14th, 21st and 35th days of growth. Analysis of *L. vetlinianus* mycelium on the 7-th, 14-th, 21-st, 35-th and 50-th days of growth was performed. *S. crispa* strains were analyzed on the 7-th, 14-th, 28-th and 60-th days. Mycelium was scraped from Petri plates and was dried for biomass evaluation.

Quenching and extraction of mycelium and fruiting bodies were carried out based on method described by Gummer with co-authors [17].

2.2 *Quenching*

Quenching of cellular metabolism was performed by 50% of cold methanol (-30°C).

2.3 Extraction

After quenching wet mycelium were ground to powder by a liquid N_2. Samples were extracted with 1.5 mL cold methanol (–25 °C) and chloroform (independently), vigorously mixed and centrifuged (10 min, 400 x g) at room temperature.

The supernatant was transferred to a new vial and the precipitated pellet was re-extracted with 1.5 mL of cold methanol. All extracts were drying by a rotary evaporator.

For a more detailed study of the accumulation of individual compounds in cultures, extraction by chloroform was applied. Crystals collected from the lids of the Petri dish and the surface of the *S. crispa* mycelium and aggregates from the *L. vetlinianus* mycelium were dissolved in chloroform and analyzed by GC–MS.

2.4 GC-MS analysis

GC–MS –based metabolomic analysis was carried out according to the standard derivatization scheme using bis—3-methyl-silyl-3-F-acetamide (BSTFA, Sigma). Silylated samples were analyzed using an Agilent 6850 gas chromatograph interfaced with 5975C mass selective detector. An HP5-MS capillary column (30 m x 0.25 mm inner diameter; film thickness of 0.25 um) was used with helium as a carrier gas at a constant rate of 1 ml/min. The temperatures of the injector and MS source were maintained at 320 °C and 230 °C, respectively. The column temperature program consisted of injection at 70 °C with an increase of 4 °C/min up to 320 °C followed by an isothermal hold at 320 °C for 15 min. Tricosane (10 ug) was used as an internal standard for the quantification of analytical results in semiquant mode, excluding sensitivity coefficients. The samples (0.5 ul) were injected in splitless mode using direct Ultra Inert Liner (Restek). The mass spectrometer was operated in the electron impact mode with an ionization energy of 70 eV. The scan mass range was set from 50 to 1000 Da at 1.27 scans per second. The data were processed and quantified with the AMDIS software (http://www.amdis.net). Compounds were identified through comparison with the retention characteristics and mass spectra of authentic standards, reported mass spectra, and the mass spectral library of the GC–MS data system (NIST 2010). The sum of the extracted ion chromatograms of the ions associated with a compound was used for quantification by UniChrom software (http://www.unichrom.com).

2.5 Statistical analysis

Statistical analysis was carried out using Microsoft Excel 2016 and Metaboanalyst.

3. Results and discussion

3.1 Growth and cultural characteristics of the fungi

D. tricolor LE-BIN 2266 grew faster than strains of the other studied species. Average growth rate on BWA – 5,5 ± 0.5 mm/d. The dynamics of biomass change is shown on the **Figure 4**. On BWA the advancing zones of colonies even, hyphae surface or raised and fringed. Colony mat wooly with long aerial interwoven hyphae around inoculum then more or less radially ordered (**Figure 5**), white, with age becoming yellowish with brown droplets of exudate and crustose hazel or brown areas. Fruiting in culture sometimes observed. Aerial hypha 1–5 μm wide, branched with regular clamps, skeletal hypha long rare branched thick walled, 2–3 μm wide (**Figure 6**). Laccase is positive.

L. vetlinianus strains showed rather slow growth rate that varied depending on strain and used medium [13, 14]. In our study, average growth rate on BWA was 1.0 ± 0.3 mm/d, 2.0 ± 0.9 mm/d and 2.9 ± 0.6 mm/d for the strains LE-BIN 2335,

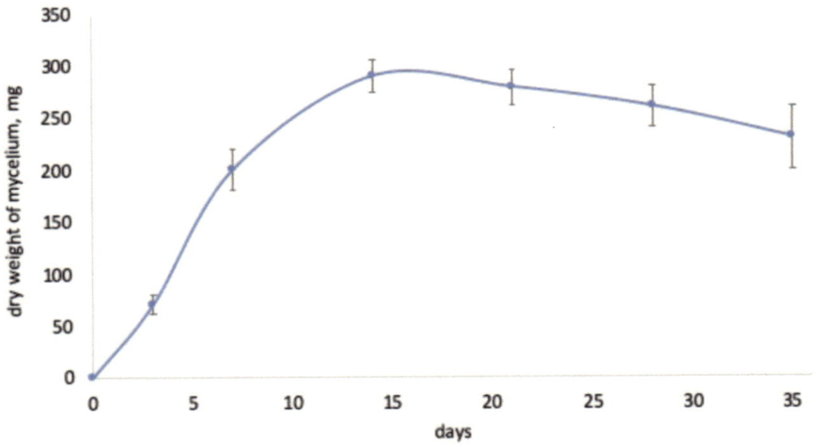

Figure 4.
Dynamic of biomass accumulation of D. tricolor *LE-BIN 2266.*

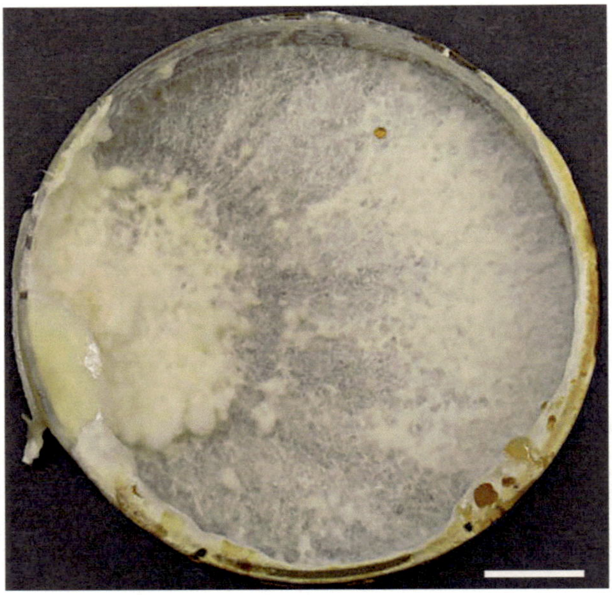

Figure 5.
Colony of D. tricolor *LE-BIN 2266 on BWA, 6 weeks. Bar = 10 mm.*

LE-BIN 2339 and LE-BIN 3253 respectively. The dynamics of biomass change is shown on the **Figure 7**. Characters of colonies were also variable depending on strain and used medium. On BWA the advancing zones of colonies even, hyphae surface or submerged. Colony mat wooly consisting of thin spread on the surface zonate aerial mycelium with more or less expressed radial fibrous hyphal bands (**Figure 8**), with age abundant fruiting [13] and conglomerates of crystals appear on the colonies (**Figures 9** and **10**). Aerial hypha monometic, 1.5–5 μm wide, branched with regular clamps (**Figure 11**). Laccase is positive.

S. crispa strains grew the most slowly in comparison with the other cultures in this study. In our experiments average growth rate of the strain LE-BIN 2902 on BWA was 0.5 ± 0.1 mm/d. The dynamics of biomass accumulation for involved in the experiment three strains is shown on the **Figure 12**.

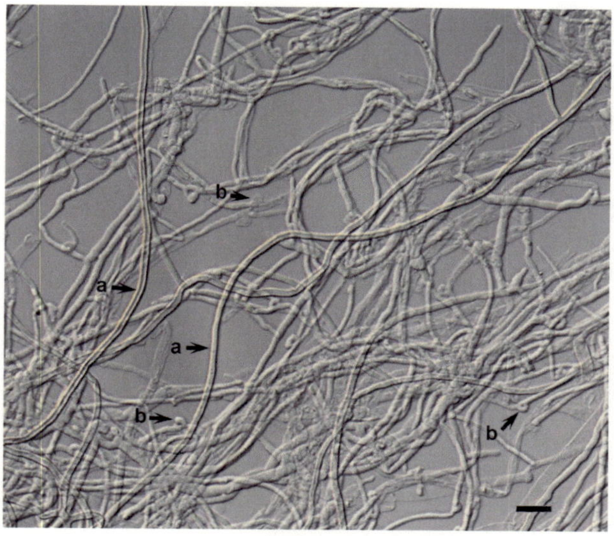

Figure 6.
Micromorphology of D. tricolor *LE-BIN 2266. Arrows indicate: a – Skeletal hypha; b – Clamp connections. Bar = 10 μm.*

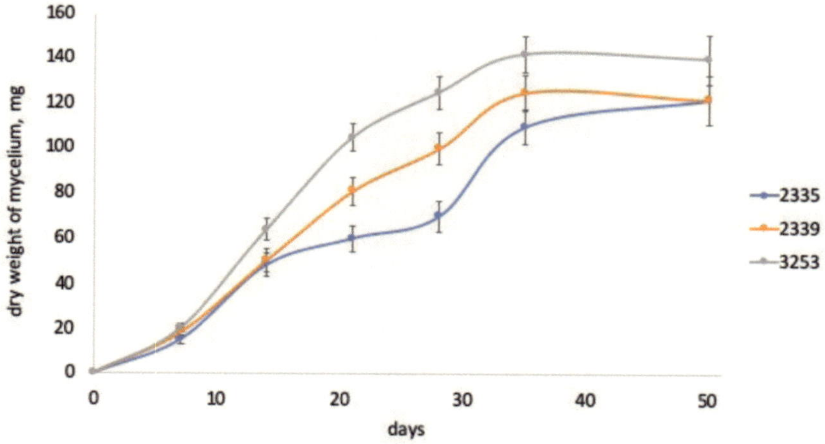

Figure 7.
Dynamic of biomass accumulation of L. vetlinianus *LE-BIN strains.*

Colony of LE-BIN 2902 on BWA is presented on **Figure 13**. The advancing zone of the colony is even, or slightly waved, marginal hyphae distant, appressed and submerged. Colony mat first downy later becoming wooly with aerial interwoven hyphae, white, creamy to grayish. Odor is absent.

Transparent crystals of various size and form can be observed on mycelium or on edges and lids of plastic plates (**Figure 14**). Over time, the number and size of such crystals increased. Aerial hypha irregular 2.0–5.0 μm wide with clamps and numerous swellings up to 10–15 μm in diam. Often arranged in chains (**Figure 15**). Single and grouped crystals were observed under the microscope. Laccase is negative.

3.2 Chemical diversity of metabolites detected by GS-MS in fungal mycelium

The results of metabolomic analysis showed that the GC–MS method allows to detect of many low molecular weight metabolites, among which are identified:

Figure 8.
Colony of L. vetlinianus *LE-BIN 2339 on BWA, 8 weeks. Bar = 10 mm.*

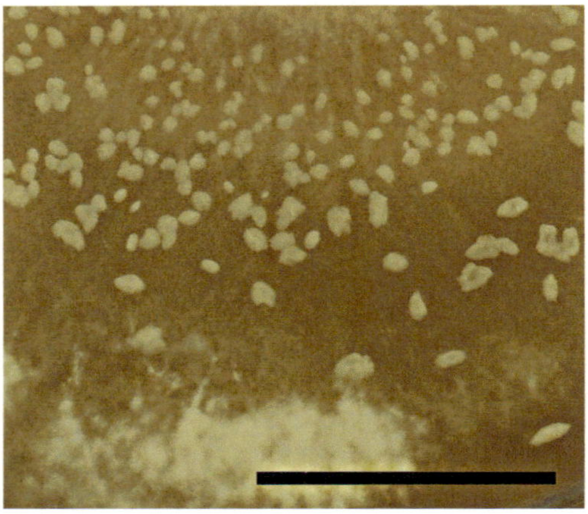

Figure 9.
Conglomerates of crystals in mycelium of L. vetlinianus *LE-BIN 2335. Bar = 10 mm.*

Figure 10.
Conglomerates of crystals in mycelium of L. vetlinianus *LE-BIN 2335 under the microscope in UV light. Bar = 10 μm.*

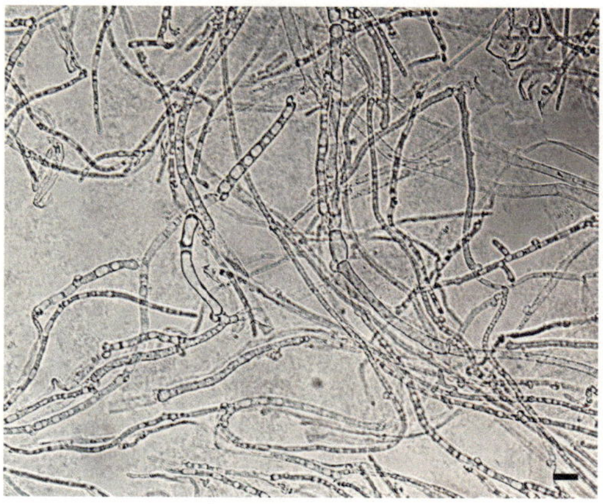

Figure 11.
Micromorphology of L. vetlinianus *LE-BIN 2335. Bar = 10 µm.*

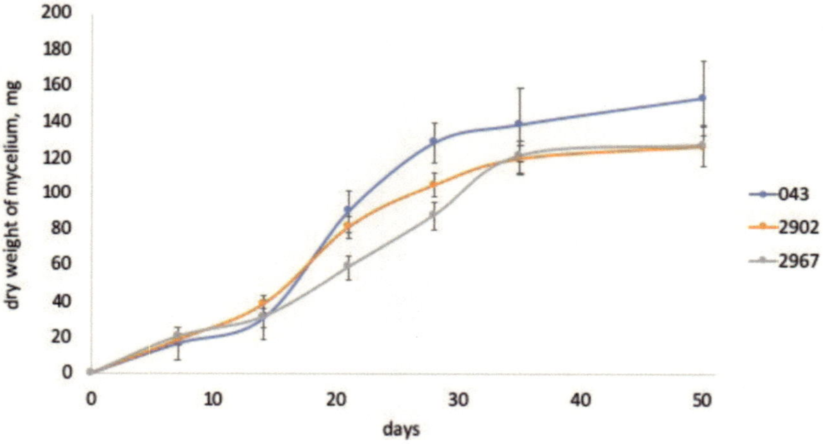

Figure 12.
Dynamic of biomass accumulation of S. crispa *LE-BIN strains.*

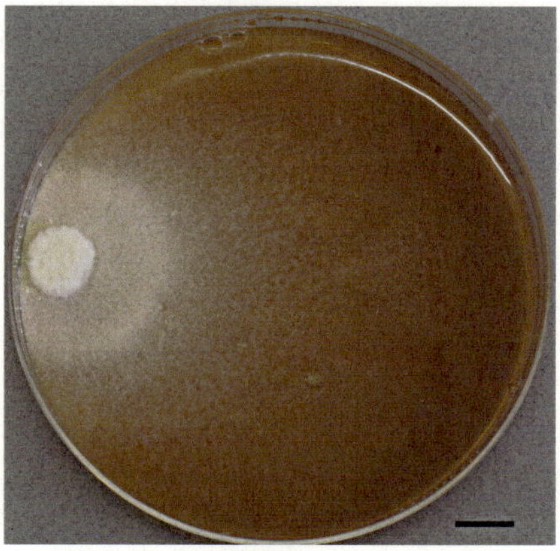

Figure 13.
Colony of S. crispa *LE-BIN 2902 on WBA, 8 weeks. Bar = 10 mm.*

amino acids, TCA (tricarboxylic acid) cycle, sugar-acids, fatty acids, monosaccharides, di- and trisaccharides, poliols, hydroxypropionic acid, cyclic metabolites, including phenolic compounds, glyceric acid, pyruvic acid, nicotinic acid,

Figure 14.
Crystals on the lid of petri plate with S. crispa. *Bar = 10 mm.*

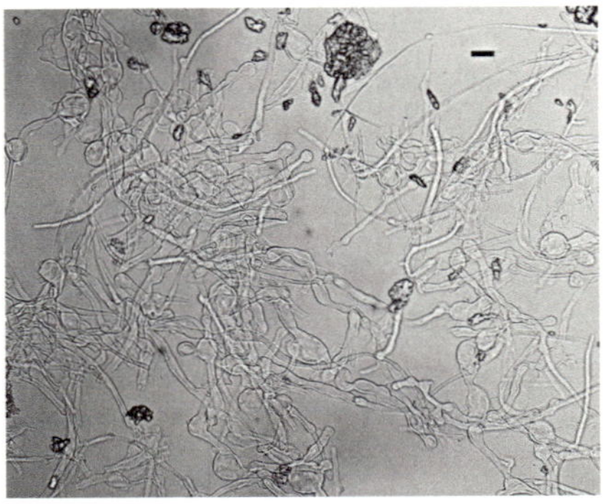

Figure 15.
Micromorphology of S. crispa *LE-BIN 2902. Bar = 10 μm.*

phosphate, methyl phosphate, glyceraldehyde, glycerol-3-P, phosphoglyceric acid (PGA).

The qualitative composition of most of the metabolites was generally similar for the three studied species of fungi. *S. crisra* mycelium contained very high quantitative of sugar that made it difficult to analyze a number of other compounds in the metabolomic profile. For this reason we do not present a detailed matrix of metabolites and statistical models for this species. The main differences between species, bur show the main differences between species only by cyclic compounds.

Specific phenolic compounds (sparassol, methyl ester of sparassol and methyl ester of orsellinic acid) were found in the *S. crispa* mycelium, as well as compounds similar to orsellinates in structure with molecular weights of 298 and 356. In some *S. crispa* cultures, on the mycelium surface and on the inner surface of Petri dish lids were found numerous clear crystals (**Figure 14**). These crystals were identified by GC–MS as methyl 2-hydroxy-4-methoxy-6-methyl benzoate (sparassol). The most intense sparassol formation was observed in the LE-BIN 2902 strain.

The mycelium of *L. vetlinianus* distinguished by the accumulation of furoic acid, cyclohexanone-3-carboxylic acid, 3-oxo-1-cyclohexene-1-carboxylic acid, and 4, 6-dimetoxy-1 (3H) -isobenzofuranone. The fastest growing strain LE-BIN 3253 also contained cyclopentene-3-carboxylic acid, in contrast to the other two *Lignomyces* strains. Numerous crystals-like aggregates glowing in UV were noticed in the plates with *L. vetlinianus* cultures (**Figures 9, 10**). They were picked up from the plates and identified by GS-MC as clusters of 4,6-dimethoxy-phthalide or 4, 6-dimetoxy-1 (3H) -isobenzofuranone, also found in mycelium.

3.3 Changes in the metabolomic profile of fungi during ontogenesis

The analysis of metabolites was carried out on the following stages of mycelium growth: the onset of growth, the formation of differentiated mycelium, the log phase, the stationary growth phase, and aging.

General patterns of changes in the metabolic network in the ontogeny of fungal mycelium were visualized using the method of principal components (PC). The dynamics of changes in the metabolomic profile of fungi had its own characteristics both at the level of the species and the strain. *D. tricolor* was characterized by the separation of the initial growth stage (day 3) and the exponential growth stage (day 7) (**Figure 16a**). All other points formed a single cluster.

A similar pattern was observed for the statistical model of the dynamics of the *L. vetlinianus* LE-BIN 2339 metabolomite network in ontogenesis (**Figure 16b**). The stages of the exponential growth phase (7, 14, and 21) were distributed in the space, and 35th and 50th days were grouped together.

The distribution of points characterizing the development of *L. vetlinianus* LE-BIN 2335 strain (**Figure 16c**) can be described as a cyclical trend, in which the

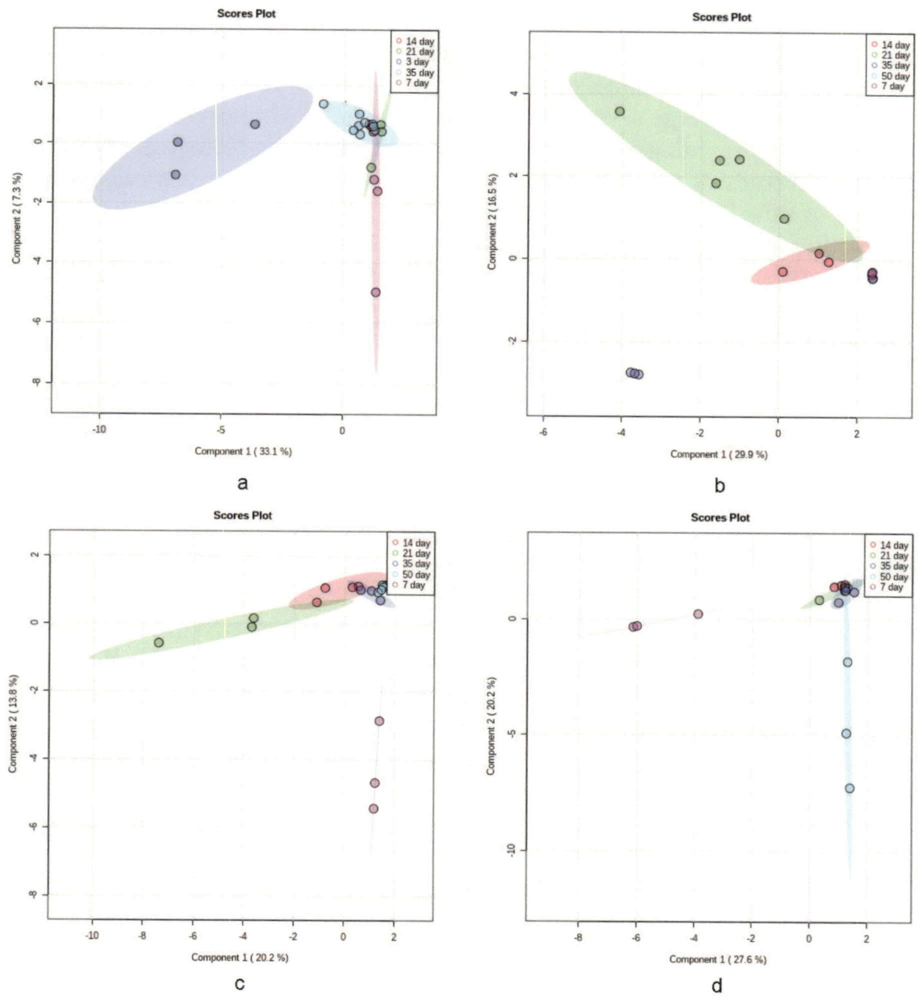

Figure 16.
Principal component analysis (PCA) of the metabolomic profile of fungi in ontogenesis: a – D. tricolor *LE-BIN 2266; b -* L. vetlinianus *LE-BIN 2339; c -* L. vetlinianus *LE-BIN 2335; d -* L. vetlinianus *LE-BIN 3253.*

7th and 50th days of growth formed a single cluster, and other stages of growth were distributed sequentially.

The distribution pattern of the points visualizing the mycelium metabolome in the fast-growing *L. vetlinianus* LE-BIN 3253 strain (**Figure 16d**) differed by an almost unexpressed discrepancy at 14, 21, and 35 days of growth. At the same time, the distribution of points describing the beginning of growth (7th day) and aging (50th days) was very different. Judging by the dynamics of changes in the mycelium biomass, the stationary phase of growth is reached only on day 35, as in other *L. vetlinianus* strains; nevertheless, the beginning of exponential growth, i.e. 7th day is distinguished by the most contrasting metabolic differences.

Thus, the isolation of the initial stage of exponential growth was general for all studied fungi. For all strains, with the exception of *L. vetlinianus* LE-BIN 3253 points characterizing the exponential growth phase on different days were distributed into separate clusters. Joint grouping of 14–21 and 35 days of growth *L. vetlinianus* LE-BIN 3253 into a single cluster is probably associated with striking changes in the metabolic network on the 7th and 50th days.

In systems biology, approaches are being developed to model the behavior of biological systems consisting of a large number of interacting variables, the dynamics of which is determined by numerous linear or non-linear relationships [18]. The method based on nonlinear time series analysis provides a global view of the dynamics of biological systems. A point in multidimensional space represents the state of a system with n-set of variables. Temporary changes in the system, formed by many points, converge to specific regions of space. The importance of this approach lies in providing a global view of the dynamics of a biological object, the experimental study of which is carried out using "-omic" approaches and assumes the factor of time [19].

According to the our results, the metabolomic analysis by GC–MS with subsequent processing of the metabolomic matrix by multivariate statistics methods makes it possible to reveal the biochemical changes of fungal mycelium during ontogenesis. The exponential phase of mycelium development is characterized by pronounced differences during of growth, which manifests itself both in the analysis of specific compounds and in the modeling of the statistical model of the metabolic network. The metabolomic network in the stationary growth phase is less susceptible to changes over time, and is also characterized by a lower dispersion of samples from one aging group.

The distribution of the main metabolites, depending on the stage of development are shown on the heat maps (**Figure 17**). At the initial stages of *D. tricolor* growth, amino acids and organic acids of the TCA cycle accumulated in the undifferentiated mycelium. Subsequently, at the stage of differentiated growth (7th - 21st days of growth), the concentrations of these compounds decreased and slightly increased on the 35th day of growth. On the 14th day of growth, the amount of glucose and some unidentified monosaccharides, as well as trisaccharides increased. On the 21st day of growth, mainly fructose, arabinose, sugar acids and some unidentified monosaccharides with retention time (Rt) 21–22 minutes accumulated. On the 35th day, the total amount of carbohydrates significantly decreased, but the accumulation of phosphate increased.

The dynamics of changes in the metabolomic profile of *L. vetlinianus* was very different from *D. tricolor*. While in *D. tricolor* the maximum concentrations of most metabolites were observed at the beginning of the exponential growth phase, in *L. vetlinianus* the concentration changes of small molecules varied greatly at all growth stages. The initial stage of *L. vetlinianus* growth was characterized by active accumulation of monosaccharides (glucose, fructose and mannose) and a number of unidentified sugars (Rt 20–22 min). Amino acids

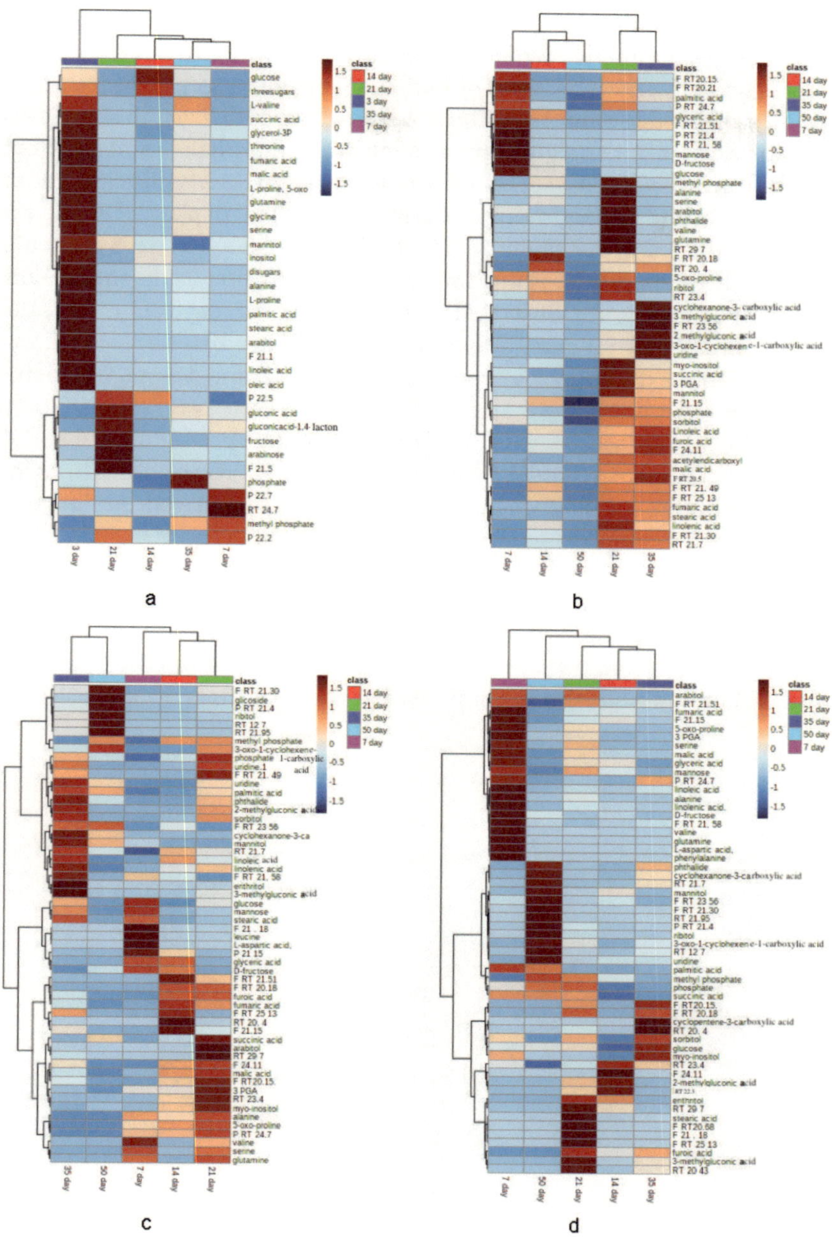

Figure 17.
Heat map of metabolites changes in the ontogenesis of fungi: a- D. tricolor *LE-BIN 2266; b -* L. vetlinianus *LE-BIN 2339; c -* L. vetlinianus *LE-BIN 2335;* d - L. vetlinianus *LE-BIN 3253 (for unidentified metabolites: Rt – Retention time, P- pyranose, F – Furanose).*

accumulated in young cultures of *L. vetlinianus* LE-BIN 2339 and LE-BIN 3253. In *L. vetlinianus* LE-BIN 2335, the maximum amino acid content was found on the 21st day of growth. In general, the strain LE-BIN 2335 with the least intensive growth was characterized by a very poor metabolomic profile on the 50th day of growth, while the faster-growing strains accumulated glycosides, sugars, and cyclic compounds on the 50th day of growth.

Cyclic compounds are mainly dominated in stationary phase of growth. But in strain *L. vetlinianus* LE-BIN 2335, the maximum content of phthalide was recorded in the mycelium on the 21st day of growth, and then it decreased. Perhaps this is due to the excretion of phthalide into the medium or to the uneven distribution of phthalide clusters on the mycelium. In *L. vetlinianus* LE-BIN 3253, which forms the smallest number of aggregates, the maximum of phthalide concentration in the mycelium was recorded on the 50th day of growth. The isolation of the 50th day into a single cluster is probably associated with a significant accumulation of phthalide and other cyclic compounds from the mycelium of this strain. Fatty acids dominated mainly at the end of the exponential and the beginning of the stationary growth phase. Di- and three-saccharides dominated in the young (7 days-old) and in aging (50 days-old) cultures.

The strain differences of *L. vetlinianus* in the composition of the metabolomic profile, described above, are illustrated on the 7th and 35th days of growth, that is, in the exponential and stationary growth phases in **Figure 18**.

In the methanol extracts of *S. crispa* amino acids (alanine, valine, isoleucine, serine, threonine, proline), sugar alcohols (glycerol, ribitol), and sugars (fructose, glucose, melibiose, sucrose), organic acids and phosphate predominated at the beginning of the log phase (day 7) of growth. On the 14th day of growth, as in *D. tricolor*, the amount of monosaccharides (Rt 18–22 min) increased. At the same time, the qualitative composition of other metabolites in the *S. crispa* mycelium practically did not differ on the 7th and 14th days of growth. With the onset of the stationary growth phase (21 days), the total amount of sugars decreased.

On the 28th day of growth, the composition of the metabolomic profile of *S. craspa* did not differ much from the three-week-old cultures, but the total quantitative content of amino acids was lower than in younger cultures by 30–35%.

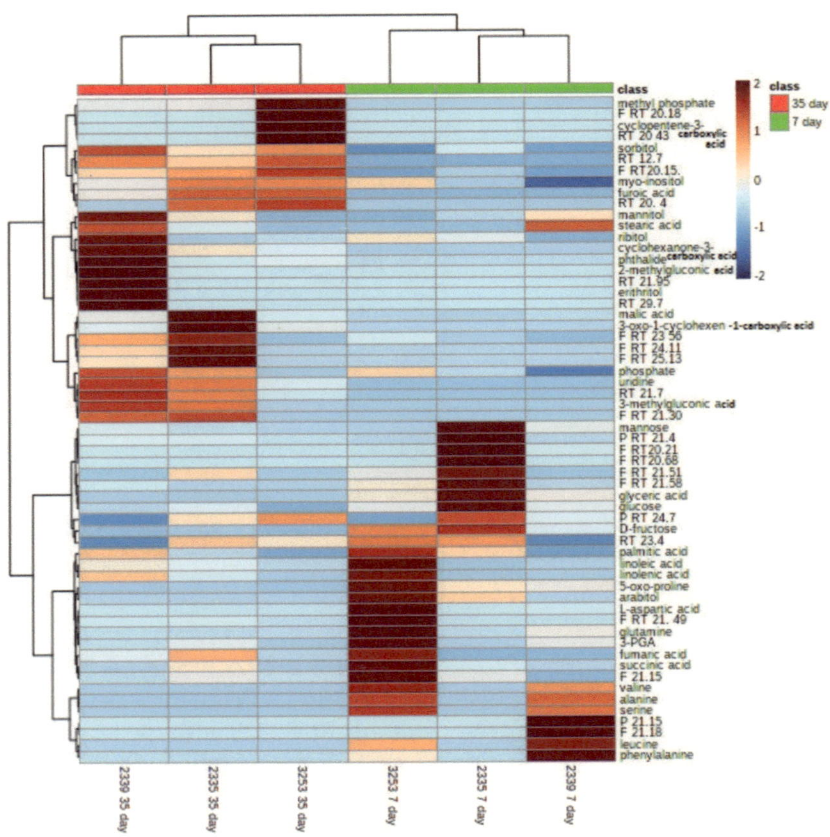

Figure 18.
Heat map of metabolites of L. vetlinianus *strains at different stages of ontogenesis (for unidentified metabolites: Rt – Retention time, P- pyranose, F – Furanose).*

DOI: http://dx.doi.org/10.5772/intechopen.96621

S. crispa strain number	Cultivation time (days)	Compounds (μg/g mycelium)		
		Methyl ester of sparassol	**Sparassol**	**Methyl ester of orsellinic acid**
		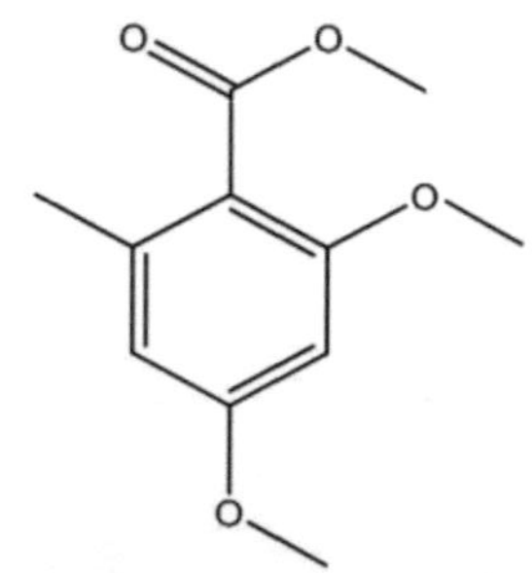	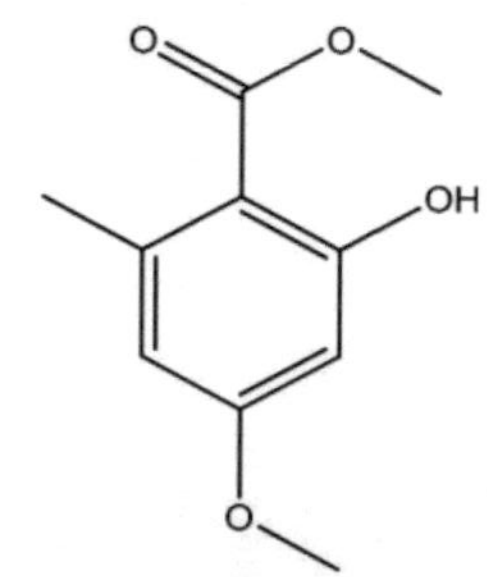	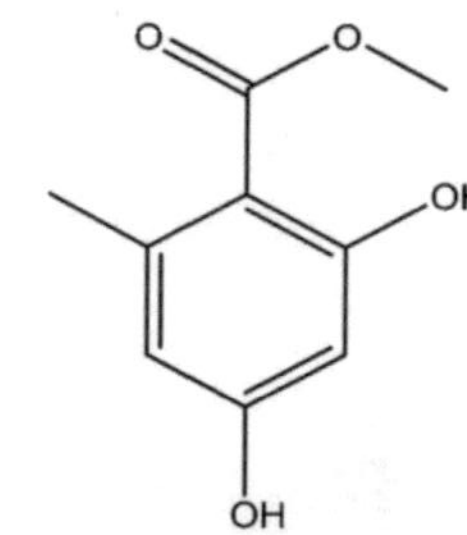
LE-BIN 043	21	—	875.0 ± 97.3	4375.5 ± 802.8
	35	509.2 ± 55.4	2598.5 ± 505.1	124.1 ± 67.5
LE-BIN 2902	21	—	905.0 ± 108.4	4013.5 ± 604.1
	35	7.6 ± 2.2	4915.7 ± 678.2	2706.3 ± 303.3
LE-BIN 2967	21	—	181.9 ± 20.9	197.6 ± 19.1
	35	17.6 ± 4.4	1638.1 ± 80.8	424.9 ± 54.3

Table 1.
Composition of phenolic compounds in the mycelium of S. crispa *strains at different cultivation periods.*

The composition of phenolic compounds synthesized by *S. crispa* and their change with the age of culture is presented in the **Table 1**. In strains 2902 and 043, methyl-orsellinate dominated in the mycelium on the 21st day of growth; its amount was 4.5–6 times higher than the amount of sparassol. In *S. crispa* LE-BIN 2967, the concentrations of these compounds were the same. On the 35th day of growth in strain 2902, the amount of sparassol increased by 4 times, the amount of methyl orsellinate decreased by 35%. In strain LE-BIN 2967, the content of sparassol increased 9.5 times, and methyl orsellinate increased 2 times. In *S. crispa* LE-BIN 043, the amount of sparassol increased by 3 times, while the amount of methyl-orsellinate, on the contrary, decreased by more than 10 times. Also, all cultures with age in the medium accumulated methyl ester of sparassol and compounds with molecular weights of 298 and 356.

Phenolic compounds found in the culture of *L. vetlinianus* and *S. crispa* are derivatives of orsellinic acid and are widespread in fungi of various taxonomic groups. In recent studies, much attention is paid to the biosynthesis, as well as the biological activity of these compounds [20, 21]. Orsellinates are believed to be involved in allelopathic interactions, exhibiting a rather weak antifungal effect [16, 22, 23]. Orsellinic acid itself is formed in the acetate-malonate pathway and is an intermediate for many phenolic compounds [24]. Many natural phthalides display a variety of biological activities. It was stated that phthalides are responsible for numerous bioactivities, however their exact mechanism of action is not studied yet [25]. Phthalides are known to be responsible for antimicrobial, antifungal, cytotoxic, enzyme inhibiting, antitumor and plant hormone like activities [26, 27]. Also there are some results suggested that phthalides acted as inhibitors of oxidative and inflammatory stress.

4. Conclusion

According to the results, the metabolomic analysis by GC–MS with subsequent processing of the metabolomic matrix by multivariate statistics methods makes it possible to reveal the species and strain specificity of biochemical changes in fungal mycelium during ontogenesis. The exponential phase of mycelium growth is characterized by pronounced differences during of growth, which manifests itself both in the analysis of specific compounds and in the modeling of the statistical model of the metabolic network. The metabolomic network in the stationary growth phase is less susceptible to changes over time, and is also characterized by a lower dispersion of samples from one aging group. High concentrations of phthalide accumulated in *L. vetlinianus* and phenolic compounds synthesized by *S. crispa* revealed in some LE-BIN strains under certain cultivation conditions allow considering these strains as promising objects for biotechnology and the study of phenolic compounds metabolism in fungi.

Acknowledgements

The cultural growth and morphology were studied within the framework of the BIN RAS research project "Biodiversity, ecology, structural and functional features of fungi and fungus-like protists" AAAA-A19-119020890079-6. The metabolomic study was supported by Russian Foundation for Basic Research № 20-04-01092 A "The mechanisms of formation and remodeling of the molecular diversity of phosphatidylcholines and their physiological role in the fungal cell". For metabolite identification, we employed databases created in the course of the implementation of the BIN RAS research project AAAA-A18-118032390136-5 "Assessment of changes in the correlation structure of metabolic networks in the process of growth and development of fungi and plants from the viewpoint of systemic biology", using equipment at The Core Facility Centre at the Komarov Botanical Institute RAS. The study of the statistical regularities of the metabolic networks of fungi was carried out within the framework of Grant of the President of the Russian Federation for state support of young Russian scientists № MK-799.2021.1.4 "The metabolomics of communities of microorganisms of lithobiontic systems".

Conflict of interest

The authors declare no conflict of interest.

Abbreviations

BSTFA	bis—3-methyl-silyl-3-F-acetamide
BWA	beer-wort agar
CE-MS	capillary electrophoresis-mass spectrometry
F	furanose
GC–MS	gas chromatography–mass spectrometry
LC–MS	liquid chromatography-mass spectrometry
MEA	malt extract agar
NMR	nuclear magnetic resonance
P	pyranose
PC	principal components
PCA	principal component analysis
PGA	phosphoglyceric acid.
Rt	retention time
TCA	tricarboxylic acid

Author details

Katerina V. Sazanova[1,2]*, Nadezhda V. Psurtseva[1] and Alexey L. Shavarda[1,3]

1 Komarov Botanical Institute RAS, Saint Petersburg, Russia

2 The Archive of the Russian Academy of Sciences, University Emb., Saint Petersburg, Russia

3 Research Park of Saint-Petersburg State University, St. Petersburg, Russia

*Address all correspondence to: barinova-kv@mail.ru

References

[1] Buchweitz LF, Yurkovich JT, Blessing C. et al. Visualizing metabolic network dynamics through time-series metabolomic data. BMC Bioinformatics. 2020;21(130). doi.org/10.1186/s12859-020-3415-z.

[2] Batygin NF. Physiology of ontogenesis. 09 In the book: Theoretical foundations of 10 selection. Vol. 2, part 1. SPb. 1995. p. 14-97. In Rus.

[3] Calvo AM, Wilson RA, Bok JW, KellerNP.RelationshipbetweenSecondary Metabolism and Fungal Development. Microbiol Mol Biol Rev. 2002;66(3):447-459. doi.org/10.1128/MMBR.66.3.447-459.2002

[4] Hegeman AD. Plant metabolomics - Meeting the analytical challenges of comprehensive metabolite analysis. Brief. Funct. Genom. 2010;9:139-148. https://doi.org/10.1093/bfgp/elp053

[5] Khakimov B, Bak S, Engelsen SB. High-throughput cereal metabolomics: Current analytical technologies, challenges and perspectives. J. Cereal Sci. 2014;59:393-418. https://doi.org/10.1016/j. jcs.2013.10.002

[6] Okazaki Y, Saito K. Recent advances of metabolomics in plant biotechnology.Plant Bio- technol. Rep. 2012;6:1-15. https://doi.org/10.1007/s11816-011-0191-2

[7] Aurand CR, Bell DS, Lamb T, Bell-Pedersen D. Metabolomic Profiling of Neurospora crassa Fungi Using HILIC and Reversed Phase LC-MS. Reporter US. 28(3). https://www.sigmaaldrich.com/technical-documents/articles/reporter-us/metabolomic-profiling.html

[8] Smedsgaard J, Nielsen J. Metabolite profiling of fungi and yeast: from phenotype to metabolome by MS and informatics. Journal of Experimental Botany. 2005;56(410): 273-286. doi.org/10.1093/jxb/eri068

[9] Kim EM, Jung HR, Min TJ. Purification, structure determination and biological activities of 20(29)-lupen-3- one from Daedaleopsis tricolor (Bull. ex Fr.) Bond. et Sing. Bull Kor Chem Soc. 2001;22(1):59-62.

[10] Teplyakova TV, Psurtseva NV, KosogovaTA,MazurkovaNA,KhaninVA, Vlasenko VA. Antiviral activity of polyporoid mushrooms from Altai Mountains. International journal of medicinal mushrooms. 2012;141:37-45. doi.org/10.1615/intjmedmushr.v14.i1.40

[11] Melzig MF, Pieper S, Siems WE, Heder G, Bottger A, Liberra K, Lindequist U. Screening of selected basidiomycetes for inhibitory activity on neutral endopeptidase (NEP) and angiotensin-converting enzyme (ACE). Pharmazie. 1996;51(7):501-503.

[12] Tomasi S, Lohezic-Le Devehat F, Sauleau P, Bezivin C, Boustie J. Cytotoxic activity of methanol extracts from Basidiomycete mushrooms on murine cancer cell lines. Pharmazie. 2004;59(4):290-293.

[13] Petersen RH, Psurtseva NV, Zmitrovich IV, Chachuła P, Arslanov S, Hughes KW. Lignomyces, a new genus of pleurotoid Agaricomycetes. Mycologia. 2015;107(5):1045-1054. doi.org/10.3852/14-355

[14] Sazanova KV, Psurtseva NV, Shavarda AL. Cultural and metabolomic studies of a new phthalids producer Lignomyces vetlinianus. International journal of medicinal mushrooms. 2018;20(11):1031-1045. doi.org/ 10.1615/IntJMedMushrooms.2018028687.

[15] Kimura T. Natural products and biological activity of the pharmacologically active cauliflower

mushroom Sparassis crispa. BioMed Research International. 2013; Article ID 982317. doi.org/10.1155/2013/982317.

[16] Woodward S, Sultan HY, Barrett DK, Pearce RB. Two new antifungal metabolites produced by Sparassis crispa in culture and in decayed trees. Journal of General Microbiology. 1993;139:153-159. doi.org/10.1099/00221287-139-1-153

[17] Gummer JP, Krill C, Du Fall L, Waters OD, Trengove RD, Oliver RP, Solomon PS. Metabolomics protocols for filamentous fungi. Methods Mol Biol. 2012;835:237-254. doi.org/10.1007/978-1-61 7 79-501-5_15

[18] Zhang B, Tian YMS, Zhang Z. Network Biology in Medicine and Beyond. Circ. Cardiovasc. Genet. 2014;7(4):536-547. doi.org/10.1161/CIRCGENETICS.113.000123

[19] Grigorov MG. Global dynamics of biological systems from time-resolved omics experiments. Systems biology. 2006;22(12):1424-1430. https://doi.org/10.1093/bioinformatics/btl119

[20] Regueira TB, Kildegaard KR, Hansen BG, Mortensen UH, Hertweck C, NielsenJ.MolecularBasisforMycophenolic Acid Biosynthesis in Penicillium brevicompactum. Applied and environmental microbiology. 2011;77:3035-3043. doi.org/ 10.1128/AEM.03015-10

[21] Schroeckh V, Scherlach K, Nützmann H-W, Shelest E, Schmidt-Heck W, Schuemann J, Martin K, Hertweck C, Brakhagea AA. Intimate bacterial–fungal interaction triggers biosynthesis of archetypal polyketides in Aspergillus nidulans. Proc Natl Acad Sci. 2009;106(34):14558-14563. doi.org/10.1073/pnas.0901870106

[22] Hughes KW, Reboredo-Segovia A, Petersen RH. Transatlantic disjunction in fleshy fungi. I. The Sparassis crispa complex // Mycological Progress. 2014. V. 13. No. 2. – P. 407-427. doi.org/10.1007/s11557-013-0927-1

[23] Peres MTLP, Mapeli AM, Faccenda O, Gomes AT, Honda NK. Allelopathic potential of orsellinic acid derivatives. Brazilian archives of biology and technology. 2009;52(4):1019-1026. doi.org/10.1590/S1516-89132009000400027.

[24] Legaz ME, de Armas R, Vicente C. Bioproduction of depsidones for pharmaceutical purposes. In: Drug development - A Case study based insight into modern strategies. Ed. by Rundfeldt C. 2011. P. 487-508. doi.org/10.5772/27051

[25] León A, Del-Ángel M, Ávila JL and Delgado G. Phthalides: distribution in nature, chemical reactivity, synthesis and biological activity. Progress in the Chemistry of Organic Natural Products. 2017;140:127-244. doi.org/10.1007/978-3-319-45618-8_2

[26] Huff T, Kuball HG, Anke T. 7-Chloro-4,6-dimethoxy-1(3H)-isobenzofuranone and basidalin: antibiotic secondary metabolites from Leucoagaricus carneifolia Gillet (basidiomycetes). Z Naturforsch C. 1994;49(7-8):407-410. doi.org/ 10.1515/znc-1994-7-803

[27] Yoshikawa K, Kokudo N, Hashimoto T, Yamamoto K, Inose T, Kimura T. Novel Phthalide Compounds from Sparassis crispa (Hanabiratake), Hanabiratakelide A—C, Exhibiting Anticancer Related Activity. Biol. Pharm. Bull. 2010;33(8):1355-1359. doi.org/10.1248/ bpb.33.1355